SpringerBriefs in Business

For further volumes:
http://www.springer.com/series/8860

Sabine E. Herlitschka

Transatlantic Science and Technology

Opportunities for Real Cooperation Between Europe and the United States

Sabine E. Herlitschka
Infineon Technologies Austria AG
Vienna
Austria
e-mail: sabine.herlitschka@infineon.com

ISSN 2191-5482 ISSN 2191-5490 (electronic)
ISBN 978-1-4614-4384-1 ISBN 978-1-4614-4385-8 (eBook)
DOI 10.1007/978-1-4614-4385-8
Springer New York Heidelberg Dordrecht London

Library of Congress Control Number: 2012940962

Printed on acid-free paper

Springer is part of Springer Science+Business Media (www.springer.com)

Review Panel

Prof. Elias G. Carayannis, Professor of Science, Technology, Innovation and Entrepreneurship, Director of Research, Science, Technology, Innovation and Entrepreneurship, European Union Research Center (EURC), School of Business, George Washington University, Washington, DC, U.S.

Prof. Marek Konarzewski, Minister Counselor Science & Technology Affairs, Embassy of the Republic of Poland, Washington DC, U.S.

Caroline Adenberger, Editor-in-Chief *bridges* Magazine, Deputy Director, Office of Science & Technology, Austrian Embassy, Washington DC, U.S.

Philipp Marxgut, Director & Attaché for Science and Technology, Office of Science & Technology, Austrian Embassy, Washington DC, U.S.

Executive Summary

Today's major societal challenges are global in nature. Some of the most pressing challenges we face are related to climate change, energy supply, efficient energy use, and global health issues. Science and Technology (S&T) plays a key role in finding solutions to these challenges. Due to their global nature, these challenges demand collaborative approaches as no individual nation or region has the intellectual capacities nor the financial resources to respond adequately and effectively by itself. Thus, international cooperation in S&T has become more imperative than ever.

Who else, if not the two major mature economic powerhouses—Europe and the U.S.—should be the ones to take the initiative to address aforementioned global challenges via enhanced S&T cooperation? Representing approximately 54 %[1]of the world's GDP in 2011, their transatlantic partnership not only provides opportunities but should also assume responsibility for tackling these global challenges, as discussed repeatedly on the occasion of the recent EU–U.S. Summits.[2]

On the other hand, one has to ask the question if in today's globalized "S&T Enterprise," the transatlantic dimension—in particular the cooperation of Europe and the United States—does it still play a major role? Just by looking at the growth rates and dynamics in emerging economies, wouldn't it be somewhat outdated to think of Europe and the U.S. as the world's drivers for science, technology, and innovation in any way?

An investigation and analysis of the current state of the transatlantic European and The U.S. partnership in S&T in the light of today' global challenges is the focus of this study. The study seeks to provide insight into the current conditions and defining dimensions of transatlantic S&T cooperation, and offers policy recommendations on how to tap the full potential of a future transatlantic S&T partnership.

[1] Hamilton DS, Quinlan JP (2011) The Transatlantic Economy 2011: Annual Survey of Jobs, Trade and Investment between the United States and Europe. Washington DC: Center for Transatlantic Relations.

[2] Overview EU–U.S. Summits: http://www.eurunion.org/eu/Table/EU–US-Summits/.

Core Questions of the Study

More specifically, core questions of this study included:

- What are the status, relevance of, and the experiences from former or current transatlantic S&T cooperation?
- Which opportunities and new mechanisms for increased effective transatlantic European–U.S. S&T cooperation could be developed under the present conditions?
- What would be options for effective joint transatlantic efforts in the light of global challenges?
- What about coordinated transatlantic "science diplomacy" efforts?

In this study "Europe" refers primarily to the European Union and its more substantial, typically strategic activities in S&T at EU level. Due to the focus on major global challenges, individual scientist-to-scientist cooperation was not specifically taken into consideration for the study. As for the activities of individual EU Member States, these were considered to some degree, although not in a comprehensive way.

While the study is focused on strategic European–American relationships, it also addresses issues of international S&T cooperation in which both Europe and the U.S. are responding (or should/could respond) to joint opportunities and/or responsibilities in other parts of the world, e.g., Africa, Asia, etc.

Methodology of the Study

This study represents a qualitative interview-based approach targeted at capturing views and perceptions of U.S. and European key stakeholders toward effective transatlantic S&T cooperation in the light of global challenges. As such, this study has drawn a multidimensional "picture" and represents a unique source of direct, first-hand information and opinions, giving ideas of what is needed and feasible in transatlantic, strategic S&T cooperation. The results, conclusions, and policy recommendations are summarized in this report.

A broad evidence base was generated: The results of this study have been derived from approximately 80 extensive personal contacts (interviews, questionnaires), more than 40 events/ conferences/ hearings related to the wider study issues, and more than 80 published reports and references, in order to expand the base for the study.

Conclusions of the Study

In summary, the following conclusions were drawn that represent some of the defining dimensions of transatlantic S&T cooperation:

Conclusion 1

The economic reality: Europe and the U.S.—biggest economic players, not perceived and appreciated as such. Europe and the U.S. are the biggest players in economic terms—still—and have the highest level of expertise to contribute in dealing with global challenges. Their economies are heavily interconnected and dependent on each other. However, this fact is not commonly recognized and appreciated as such. The current financial and economic crisis on both sides of the Atlantic demonstrates the interconnectedness as well as the need for coordinated actions. Instead it seems to trigger the trend towards more inward-orientation and inward-looking–both in Europe and in the U.S. Particularly when it comes to the transformation to what is typically referred to as "knowledge economy", Europe and the U.S. could clearly benefit from each others' experiences and efforts, since the "knowledge economy" poses huge challenges particularly to highly developed economies with its implications in terms of future job profiles, the practical relevance of "open innovation" and the innovation system, the future role of established capabilities such as manufacturing, and the role of science, technology and innovation.

Conclusion 2

Complexity of systems perceived likewise, adds complexity at instruments level and creates unattractive framework for cooperation. Each respective system, Europe as well as the U.S. (and other systems), is perceived by the other side as highly complex and hard to understand. Repeatedly, key issues such as the variety of actors, decentralization versus coordination, specific politically-motivated interest groups, short-term interests, etc. have been mentioned. This complexity of each system is also reflected by the instruments set up towards transatlantic S&T cooperation, thus adding up to multiple layers of complexity. For participants or potential participants in transatlantic S&T cooperation this situation generates highly unattractive framework conditions.

Conclusion 3

Diverging driving interests and priorities rather separate than unify Europe and the U.S., despite pressing global challenges. Despite similar societal values, Europe and the U.S. are driven by fundamentally diverging forces: competition versus cooperation, short-term versus long-term orientation, knowledge generation versus business generation, differing views on the relevance of security, etc. Nevertheless, Europe and the U.S. are confronted with joint global challenges too big to deal with independently. And both systems are massively interconnected also in succeeding or failing when it comes to dealing with our joint global challenges.

Conclusion 4

Shared values have not yet lead to comprehensive joint visions and coordinated strategies. There are a number of issues on which Europe and the U.S. work together, based on common interests and an expedient alliance type of relationship. Europe and the U.S. are those regions closest in terms of shared values. In the light of major global challenges, a joint vision, real joint ambition and corresponding strategies based on shared values and appreciation—though important—are in question or not existing.

Conclusion 5

"typical…": stereotypes, misperceptions and prejudices still exist. Typical, simplistic stereotypes exist and are found in the public as well as in wide parts of the S&T community. Examples such as "Europe is just complicated", "with Europeans you always get lost in procedures", "Europeans are busy with themselves and their internal procedures", as well as "Americans are simplistic", "Americans are interested in short-term opportunities only", "America is driven by lobbyists' interests instead of concentrating on real issues", etc. are surprisingly wide spread.

Conclusion 6

International orientation and the "hype" for China: very different approaches. The perception of the purpose and benefits of real international S&T cooperation is fundamentally different on both sides. Also in the case of international cooperation, it is true that money follows strategy, assuming that where there is no money there is no interest. The U.S. currently seems to be driven primarily by the "hype" for China, which has to be seen in the context of both, the economic dimension and the huge U.S. financial dependence on China, as well as the sometimes naively perceived short-term business opportunities in China. Europe, so far less interconnected and dependent on China seems to implement more long-term strategies in international S&T cooperation. Particularly the EU-Research Framework Program represents a long-term financial commitment to S&T cooperation, including and providing funding for international cooperation outside Europe such as partners in China.

Conclusion 7

Experiences in strategic transatlantic cooperation: existing mechanisms are unsatisfying, new approaches are needed and are on the way, respectively. Many key stakeholders in Europe and the U.S. are either unaware of the need for strategic transatlantic S&T cooperation or unsatisfied with the way strategic transatlantic S&T cooperation is currently organized. The traditionally applied procedures for transatlantic cooperation particularly in areas of strategic importance no longer seem to fit in terms of approach, timing, content and dynamic. Many of the same hurdles still exist as many years before. Effective ways of strategic cooperation need effective mechanisms to make the cooperation work and deliver results. This clearly has to go beyond the majority of established routines. Some new initiatives are on the way.

Policy Recommendations of the Study

Based on the findings and overall conclusions, the three policy recommendations have been developed and are summarized as follows:

- Where to go together: Real Leadership is needed
- Bold actions needed: Think big and pragmatic
- Think global: Europe and the U.S. towards international cooperation

Each of the policy recommendations has been split into a recommendation for strategic development and the respective recommendation for action in order to provide guidance for implementation.

Recommendation 1: Where to go Together: Real Leadership Needed!

Recommendation for Strategic Development

The relationship between Europe and the U.S. is built on shared values and similar societal models. From the perspective of Europe or the U.S. likewise, no other region/country is closer with respect to fundamental societal values. Thus, the nature of the transatlantic cooperation is and should be more than just an expedient alliance, but an alliance of joint societal values. It requires active dedication, the agreement on a clear and committing vision for shaping future developments that go beyond short-term advantages. Joint ambitious objectives should be defined including deviated strategies plus explicit tangible results to be achieved. In essence, what is needed is the further transition of this transatlantic alliance originally set up AGAINST a joint enemy, towards an alliance FOR the development of joint solutions. The Atlantic Basin Initiative[3] can serve as good example.

Recommendation for Action

The economic and financial crisis could be instrumental in concentrating on developing a joint vision implemented by coherent strategies. This is a clear opportunity driven by importance and urgency.

Activities under the Transatlantic Business Dialogue (TABD) and the Transatlantic Economic Council (TEC) have gained substantial momentum. Specifically

---

[3] European Commission (2011) The Atlantic Geopolitical Space: common opportunities and challenges.

strategic S&T should play a major role in order to foster innovation. S&T as part of the wider innovation/competition related efforts are a corner stone of what is typically referred to as "knowledge economy" and is a developmental challenge both for Europe as well as the U.S. Close links between TABD, TEC, related activities and S&T should be actively used and expanded.

Transatlantic S&T cooperation does not take place in isolation but needs to be an integrated part of a wider system. It is shaped by the various actors, networks, structures and dynamics in the system. The understanding of transatlantic S&T as a common "system" with its interdependencies, driving forces and opportunities should be strengthened, thus facilitating the set up of a more effective common transatlantic S&T system with related mechanisms and instruments.

Recommendation 2: Bold Actions Needed: Think Big and Pragmatic!

Recommendation for Strategic Development

Time has come for bold AND pragmatic actions between Europe and the U.S. if global challenges are to be dealt with effectively. Therefore, focus is needed on major issues of concern and pressing global challenges.

Recommendation for Action:

Sustainable energy supply and energy use are amongst the top priorities, for which bold plans and implementing measures should be developed including all relevant actors. Timely and pragmatic ways towards effective solutions and achieving real results should be developed.

Fields/themes of activities should be aligned across all relevant policy areas (in the EU for instance including all responsible directorates general, such as enterprise, trade, internal/security, external, energy, etc.) and "routine" procedure of cooperation towards tangible results and delivery of effective mechanisms should be developed.

The direct involvement of scientists and engineers is an essential element in providing the passion and pressure for results. This factor should be reinforced.

Systematic analysis of significant examples of joint transatlantic S&T cooperation as well as the identification of successful examples is highly recommended in order to foster structured learning. Some new initiatives are on the way and should be checked explicitly for the effectiveness of their mechanisms for implementation.

Information is key, information on opportunities of transatlantic European–U.S. S&T cooperation should be made available more broadly, preferably with the help of established, highly renown, thematically oriented networks in Europe and in the U.S.

Recommendation 3: Think Global: Europe and the U.S. Towards International Cooperation

Recommendation for Strategic Development:

In effectively dealing with global challenges, Europe and the U.S. should include the international dimension, thus working towards developing a truly global S&T cooperation and funding instrument.

Recommendation for Action

This new global program would have to build on contributions from all countries participating in terms of content, mechanisms and financing. The EU-Framework Program for Research, Technological Development & Demonstration can serve as example, as it represents the largest transnational cooperative and competitive research program worldwide, open to the entire world. These efforts towards international cooperation should be complemented with coordinated activities across all policy areas, in particular external relations.

Science diplomacy is an effective way of fostering international cooperation. Opportunities for coordinated science diplomacy measures resulting from trans-atlantic initiatives should be systematically identified and analyzed for implementation

Forewords and the Wider Horizon of Transatlantic Relations Between Europe and the U.S.

Summary

Forewords by the Editor of this Springer book Series On SpringerBriefs in Business, and an expert in transatlantic cultural diplomacy provide the relevance and the wider context for this report.

On purpose culture has been taken as an example. Culture has become a major political and philosophical issue. Given their political and strategic importance, the so-called 'geo-cultural' issues have been called upon to constitute, along with geo-political and economic issues, a governance axis, representing aspects similar to science and technology. At a time where the EU and the U.S., despite the recession, remain each other's most important markets and face many common global challenges, they should rethink their conception and political implication of culture and establish a cultural and scientific transatlantic strategy and partnership.

Transatlantic Science and Technology Cooperation as Part of the Innovation, Technology, and Knowledge Management Book Series

Elias G. Carayannis, PhD, MBA, BScEE, CPMMA

Professor of Science, Technology, Innovation, and Entrepreneurship
Director of Research, Science, Technology, Innovation, and Entrepreneurship, European Union Research Center (EURC)
Co-founder and Co-Director, Global and Entrepreneurial Finance Research Institute (GEFRI), Department of Information Systems and Technology Management,
School of Business, George Washington University, Washington, DC 20052, Email: caraye@gwu.edu
Editor of the Book Series Innovation, Technology and Knowledge Management

This is a very timely and seminal piece of work reflecting on the status, dynamics, and trends of both knowledge creation and innovation policies and practices in Europe and the U.S. as well as the implications of the quality and efficacy of collaborative Euro–American schemes for the standard of living as well as the overall way of life of people in the Euro–American space over the long run.

The findings and insights derived from this work could and should play an important role in the emerging policies and practices as well as the prevailing views as to the role and influence of knowledge economies and societies in the non-G 7 context (especially Brazil, Russia, India, China, commonly referred to as BRICs countries) and over the next 20–30 years. This may well imply that the currently presumed "growth leaders" (the BRICs) may end up following a very different trajectory over the next 20–30 years that is expected today and the example of Japan over the last 30 years should be a not-forgotten case-in-point.

In particular, geo-political, geo-economic, geo-strategic, and geo-technological (GEO-PEST)[4,5] influence, leadership, and dominance matters will be increasingly determined not by who can be the lowest-cost manufacturer but who can develop and sustain highest-value-added competence as an innovation leader locally and globally.

Moreover, the ways and means that democracy, innovation, and development along the political, economic, strategic, and technological frontier intertwine and shape their respective evolutionary paths, will prove key factors and sources of competitive and even comparative advantage in the context of knowledge-based as well as knowledge-driven societies and economies and their underlying sectors of government, university, industry, and civil society (for more see Quadruple Innovation Helix and Mode 3 Knowledge Production Systems[6]).

Therefore, the reader of this work should not only approach it as a descriptive treatise of the way things are but also use it as an intellectual springboard to envision how things are also being and becoming in particular in the context of Schumpeterian creative destruction.

In short, tomorrow is shaped not only by what took place yesterday but also by what we can start making happen today and in particular with regards to triggering, catalyzing, and accelerating high quality and quantity innovations and especially discontinuous ones with technological disruption potential.

The listing of the key conclusions below represents important dimensions of transatlantic S&T cooperation, which should be the point of departure for this "Gedankenexperiment":

---

[4] Carayannis EG (2010) Cyber Defense and Cyber Democracy. NATO War College Seminar, Thessaloniki, Greece, December 2010.

[5] Carayannis EG (2011) Innovation Diplomacy and the Hellenic-American Innovation Bridge. Springer Journal of the Knowledge Economy, September 2011.

[6] Carayannis EG, Campbell D (2009) 'Mode 3' and 'Quadruple Helix': Toward a 21st century fractal innovation ecosystem. Int. J. Technol. Manage. 46, no. 3–4: 201–234.

- The economic reality: Europe and the U.S.—biggest economic players, not perceived and appreciated as such.
- Complexity of systems perceived likewise, adds complexity at instruments level and creates unattractive framework for cooperation.
- Diverging driving interests and priorities rather separate than unify Europe and the U.S., despite pressing global challenges.
- Shared values have not yet lead to comprehensive joint visions and coordinated strategies.
- "typical…": stereotypes, misperceptions, and prejudices still exist.
- International orientation and the "hype" for China: very different approaches.
- Experiences in strategic transatlantic cooperation: existing mechanisms are unsatisfying, new approaches are needed and are on the way, respectively.

The Wider Horizon: In Favor of a New Coordinated Transatlantic Cultural Strategy

Aude Jehan

French Embassy Fellow and Visiting Research Associate at the Center for Transatlantic Relations,
SAIS, Johns Hopkins University,
Washington DC, USA

Introduction

Over the centuries the term ‘culture’ has been invested with multiple meanings evolving with history and social changes, to the point of encompassing everything and meaning nothing.

As sociologists have shown, throughout history, cultures have evolved different rules that govern interaction in groups and conduct our everyday social lives. Culture has become a major determinant of how people perceive each other and deal with their differences. Culture is a dynamic, symbolically based, and learned system. It forms the mechanism through which people construct and enact meaning. It is a learned system of meanings, communicated by natural language and symbols that allows groups of people to manage social and physical diversity and to adapt successfully to their environment. It does this by enabling members of a social group to construct a particular sense of reality. Based on this image of the world, people (1) base expectations about what motivates others; (2) learn the “correct” way of responding to challenges in their environment; and (3) develop emotional responses to their experiences. In brief, peoples’ representational, directive, and affective frames of reference for dealing with the world around them are based in their cultural experiences. Cultural models provide a coherent, systematic arrangement for the knowledge that characterizes each cultural group.[7]

When referring to culture in this book, I have taken a broad view of what the term includes, and discuss science, sport, and popular culture as well as the performing and visual arts and heritage.[8] Rather than establishing a definition of culture the objective here is to use culture as an analytic concept. I am not

---

[7] This brief definition of culture theorized by Rubinstein (Rubinstein RA (2003) Cross-Cultural Considerations in Complex Peace Operations. Negotiation Journal 19: 29–49) summarizes material from anthropology and organizational development studies. For fuller descriptions of these materials, see Alvesson M and Berg PO (1992); Avruch K (1998); Holland N and Quinn D (1987); Jacquin-Berdal D, Oros A, and Verweij M (1998); Rubinstein RA (1989, 1993); and Shore B (1996).

[8] Accordingly to the United Nations’ 1948 Universal Declaration of Human Rights, stating that: “Everyone has the right freely to participate in the cultural life of the community, to enjoy the arts, and to share in scientific advancement and its benefits.”.

interested in the content of a "culture" but how people mobilize and use the concept for political and/or economic means. Culture has been used and abused to become a political tool, especially in Foreign Affairs. My goal, therefore, is to paint a picture that reflects the U.S.'s and the EU's conception of culture and cultural diplomacy. Over the last 50 years, "the cultural sector grew to have various roles and uses (sometimes even contradictory ones), rendering innovative and new dimensions to the social, economic or political spheres." Thus, cultural, religious, and ethnic factors have been playing a larger part in defining our sense of self and community. Nevertheless, in Foreign affairs, despite the ubiquity of culture in international relations, its importance is not well recognized or even worth, it seems "suspect".

Through a brief summary of the historical context, I examine the role of cultural diplomacy and the reasons why it is (still) a critical issue for the EU and the U.S. Finally, I advocate a new approach to cultural diplomacy, based on a new transatlantic cooperation.

Talking About Cultural Diplomacy and Exchanges

Cultural diplomacy was pioneered by the U.S. in the late 1930s in order to combat Nazi propaganda, but it has gained in significance as the world has moved from the bipolarity of the Cold War to the uncertainties of the present multipolar world. It refers to a domain of diplomacy concerned with establishing, developing, and sustaining relations with foreign states by way of culture, art, and education.[9] It is also a proactive process of external projection in which a nation's institutions, value system, and unique cultural personality are promoted at a bilateral and multilateral level. But first and foremost, cultural diplomacy refers to psychology, mentality and way of life, customs, traditions, and history. Its success is based on intercultural dialogue and mutual respect with the target nation(s) and the recognition of its own cultural specificity.

From a historical point of view, "when the United States assumed the mantle of global leadership after World War II, cultural diplomacy was considered a central part of its strategy. Thus the CIA covertly supported cultural activities abroad, organizing foreign conferences and funding intellectual publications such as *Encounter and Preuves*. These activities continued into the 1950 s, under the auspices of the newly created U.S. Information Agency (USIA)."[10] However, the end of the Cold War and the fall of the Berlin Wall brought about a radical change in the approach to culture. The apparition of a number of new independent states and the cultural justification for their independence on the international arena

---

[9] For more details about cultural diplomacy during the Cold War, see Arndt RT (2005) The first resort of kings: American cultural diplomacy in the twentieth century. Potomac books.

[10] Finn H (2003) The Case for Cultural Diplomacy: Engaging Foreign Audiences. In: Foreign Affairs Volume Vol. 82, No. 6, November/December 2003.

became a major political issue, placing culture in the heart of the debate. The concept of culture was expanded to encompass that of 'identity' itself".[11] Subsequently the notion of culture, attached to the idea of endogenous development, acquired new political substance. The link between culture and development contributed to arguments in favor of financial and administrative aid to developing countries who claimed their right to define their 'own' ways of development in order to fully and equally participate in international affairs.

Both in the U.S. and in the EU, culture and cultural exchange have become regarded as being desirable, but not essential in the field of Foreign Affairs and public diplomacy. A common view is that, while cultural diplomacy can help to establish and support working relationships between countries, it is strictly subordinate to the harder stuff of laws and treaties, bilateral negotiations, multilateral structures, and military capability.

In the U.S., since the backlash against cultural diplomacy[12] during the last 10 years, it has weakened considerably and any reference to the term itself seems even to have been avoided. Its role was deprioritized, and its potential as the centerpiece of a real American re-engagement in dialogue with the rest of the world was misunderstood. A quick overview of the cultural-exchange landscape illustrates the way in which cultural purposes were spread across government: "USAID scholarships for foreign students to U.S. institutions, for example, declined from 20,000 in 1980 to just 900 in recent years. Funding for educational and cultural exchange fell more than one-third in inflation-adjusted dollars between 1993 and 2001, from $349 million to $232 million. Academic and cultural exchanges declined from 45,000 in 1995 to 29,000 in 2001. Total funding for the Bureau of Educational and Cultural Affairs amounts to just $3 million. Congress and the White House have begun to address the funding shortfall, increasing the total funding available for public diplomacy to $1.5 billion in 2005, its highest level ever."[13] Because *culture*, in general, exists in a policy vacuum within the U.S. system,[14] any program of cultural diplomacy or exchange must overcome a number of unique obstacles. Despite an acknowledgement at the highest levels of government, the low public-policy priority generally afforded culture, combined with a multi-actor, highly privatized system of cultural production and exchange, has made it difficult for the U.S. to mount a coherent, large-scale cultural diplomacy effort tailored to contemporary challenges.

---

[11] Jehan A (2011) Culture as a key factor within Western societies and a political tool for the European Union. In: EU-Topias Journal, Vols. 1–2 2011, November 2011.

[12] Many artists argued they were not activist and don't want to take part in any form of diplomacy, worried to make culture playing the same role as during the Cold War. For more details, see Brown J (2009) The backlash against cultural diplomacy. The Huffington Post, November 8, 2009.

[13] The Curb Center, Vanderbilt University (2008) Cultural Diplomacy and the National Interest: In Search of a 21st Century Perspective. In: Arts Industries Policy Forum (AIPF) Report.

[14] The U.S. has neither a Department of Cultural Affairs nor a Ministry of Culture.

However, it is true that since 9/11, the U.S. channels of cultural communication have received increased funding and attentions but still not enough yet, leading experts to conclude: "given the pressure for immediate, measurable results on specific policy issues, any policy of cultural exchange—burdened by assumptions of give-and-take, mutual learning, and creative processes that rarely register in exit polling- stands at a significant disadvantage in the constant attention and funding."[15] Nevertheless, analyzing the official publications and data, *"exchange programs"*, defined as *"one of the most effective means of increasing mutual understanding is through people-to-people"*[16], remain one of the most important U.S. governments' action in terms of Cultural Diplomacy in the last decade. The State Department annually sponsors more than 40,000 educational and cultural exchanges—including visitors to the U.S. and Americans traveling abroad. "These exchanges offer first-hand experiences of American society and culture to foreign visitors and provide opportunities for Americans to learn about other countries, cultures and peoples. Such intercultural experiences personify the universal values of human rights, freedom, equality, and opportunity that all civilized nations share.[17]"

In the EU, exchange programs are an important tool to enhance the cultural exchange within Europe. The Sixth Framework Program's Human Resources and Mobility (HRM) activity has a budget of €1,580 million and is largely based on the financing of training and mobility activities for researchers. Under these activities, known as the Marie Curie Actions, the EU offered 4,5 billion Euros for mobility and exchanges. The Marie Curie fellowship are aimed at the development and transfer of research competencies, the consolidation and widening of researchers' career prospects, and the promotion of excellence in European research.[18] Furthermore, the Erasmus Program, launched to facilitate a mutual understanding and a deeper European integration through educational exchanges between any EU's Member State, has been one of the greatest success of the European Union. In the academic year 2008/09, a total of 198,600 students went to one of the 31 countries participating in the Erasmus program (EU Member States, Iceland, Liechtenstein, Norway and Turkey). However, lower budget increases in the next few years meant the program could not keep expanding at similar rates without additional resources.

Then, in 2009, a new cooperation and mobility program in the field of higher education, Erasmus Mundus was implemented. It "aims to enhance quality in higher education through scholarships and academic cooperation between Europe and the rest of the world and promotes dialogue and understanding between people and cultures through cooperation with Third-Countries. In addition, it contributes to the development of human resources and the international cooperation capacity

[15] The Curb Center, Vanderbilt University (2008) Cultural Diplomacy and the National Interest: In Search of a Twenty first Century Perspective. In: Arts Industries Policy Forum (AIPF) Report.

[16] Bureau of Public Affairs, U.S. Department of State (2008) Diplomacy: The U.S. Department of State at work. DOS Publication.

[17] Bureau of Public Affairs, U.S. Department of State (2008) Diplomacy: The U.S. Department of State at work. DOS Publication.

[18] See http://ec.europa.eu/research/fp6/mariecurie-actions/action/level_en.html

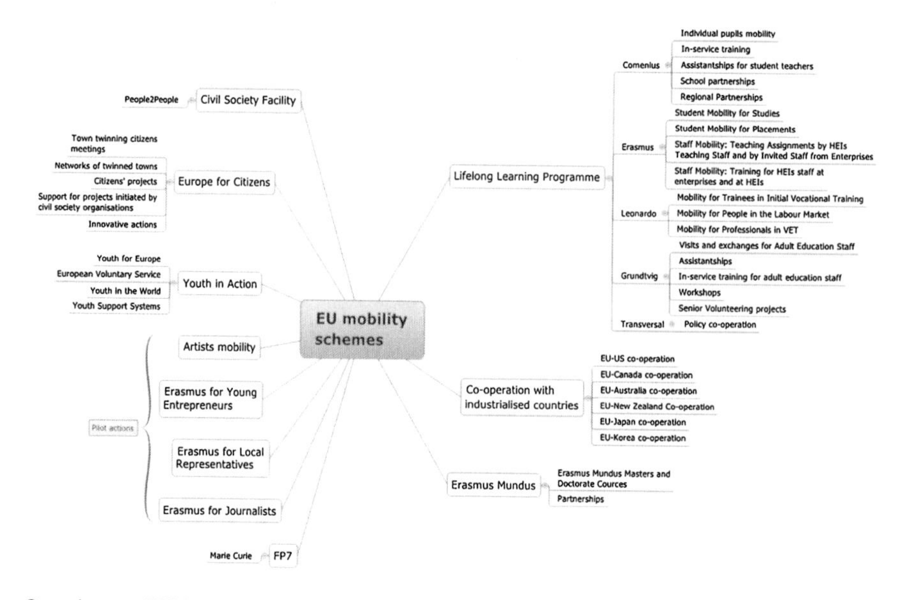

Overview on EU Mobility Schemes. *Source* Study on maximizing the potential of mobility in Building European Identity and Promoting Civic participation, for the European Commission's Education, Audiovisual and Culture Agency and DG Communication, Final Report, ECORYS, July 2011

of Higher education institutions in Third Countries by increasing mobility between the European Union and these countries."[19]

Nevertheless, as Fig. 1 shows, it represents only a very small amount from the European mobility programs, much more focused on transnational borders exchanges. Except the Fulbright Program and some agreements signed with the U.S. bilaterally by some Member States[20] and some universities (which remain local or national initiatives even though they receive financial funding from the EU[21]), it is the first educational exchange program at a Transatlantic level involving the EU as an actor and not only as a sponsor.

At a time where the EU and the U.S., despite the recession, remain each other's most important markets and face many common global challenges, they should establish a cultural and scientific transatlantic strategy and educational, cultural, or

[19] See the European Commission's website: http://eacea.ec.europa.eu/erasmus_mundus/programme/about_erasmus_mundus_en.php.

[20] Many of the individual EU Member States have invested huge amounts for various mobility and exchange programs.

[21] As an example, every year the Executive Agency publishes Compendia of all projects selected for funding under the annual call for proposals of the Lifelong Learning Program (i.e. actions managed by the Education, Audiovisual and Culture Executive Agency: Multilateral Projects, Networks, and Accompanying Measures) of the Erasmus sub-program: http://ec.europa.eu/education/erasmus/doc/compendium2010_en.pdf.

scientific exchanges should be a priority. They foster innovation's ground, straiten mutual objectives, and give us the chance to appreciate points of commonality and, where there are differences, to understand and respect them. Awareness of cultural differences, in hand with a deepened cultural self-awareness, provides a pathway to recognizing and valuing differences, and carries the potential for fruitful further learning. "*In the long course of history, having people understand your thought is much greater security than another submarine*" said J. William Fulbright. Interculturalists emphasize that there are no 'universals' that enable automatic understanding, and we must assume difference. It is important to remember that most of the time; it is not the differences between cultures that leads to conflicts and misunderstanding, but the misunderstanding and not-knowing about cultures.

The ability to mobilize '*culture*' is more than ever a precious resource in international relations, and not one that rests only in the hands of our diplomats: we all need and have a duty to realize its potential. Cultural diplomacy is certainly not the answer to any problem we face but it has a critical role to play. "As our nation and world confront the multiple challenges of our time, effective communication is imperative. Bypassing written and spoken words, art expresses the human spirit and human creativity, connecting all citizens on a deeper level," noted Secretary of State Hillary Clinton.[22]

In Favor of a New Approach of Cultural Diplomacy: A Transatlantic Cooperation

As identity politics exert an increasing influence on domestic and international exchanges, culture and cultural diplomacy could be particularly useful with emerging powers, such as India and China, placing increasing emphasis on culture in their approaches to public diplomacy.

Some issues may require a national approach but it is time to tackle the universal challenge of finding an effective balance in the relationship between culture and politics. Many of the challenges we face, such as climate change, terrorism, and managing migration, are common to the U.S. and the EU and cannot be solved by military might or unilateral policy innovations. We must create more mechanisms for engaging cultural institutions and professionals in the policy-making process so that we do not miss important opportunities. We must coordinate our efforts. Our research highlights a wealth of examples of good practice, but it suggests that the U.S. and the EU need a more strategic and systematic approach to cultural diplomacy. As Bound, Briggs, Holden and Jones already explained in 2007, "on a more fundamental level, the internet is also changing the nature of culture and the nature of the culture that we consume, and far from replacing actual experience, virtual

[22] Secretary of State Hilary Clinton (11 May 2009) Letter to the Foundation for Art and Preservation in Embassies (FAPE). FAPE was established with the objective of promoting and upholding the hallmarks of American culture throughout the world with installations at U.S. embassies in other countries.

engagement has proved a stimulus to physical participation. (…) People now expect not just to be able to access culture virtually, but also want the opportunity to add their own opinion. This represents more than personalization of choice: it means that individuals can shape and share the meaning of culture." This is the reason why, we assess that technological, scientific, and traditional (artistic) cultures should be part of a broad cultural cooperation approach, providing better understanding of the context and evolution of civilization.

For instance, cultural networks, as well as scientific networks, enable flexible ways of cooperation. They provide the opportunity to solve concrete problems that members are facing, structure for professional virtual communities in cultural sector, and efficient communication channels for their members. To enlarge them at a transatlantic level (via a common data-based for instance) could be very successful and quite easy to implement.

This realization could even be considered as a concrete challenge for the *Transatlantic Legislator's Dialogue*, launched in 1999 between the European Parliament and the American Congress. More specifically, we identified three different topics which are, in our opinion, of major interest for such a new coordinated transatlantic approach and could be part of a transatlantic agreement in the framework of TLD:

First and foremost, ensuring *a balanced support for cultural exchanges* would be crucial. Based on and with the support of the current Fulbright Program, new transatlantic exchange cooperation for cultural professionals, scientists, scholars, and students could be launched, encouraging a sense of mutation from the exclusively Western system of reference to a broader one, relating South and East, East and North, East and East, etc. Instead of promoting only bilateral exchanges between the U.S. and other countries like U.S. France or U.S. Germany and U.S. China (which is the current Fulbright Program's format), this new partnership would involve both the EU and the U.S. with the financial support of the U.S. and of the EU, and promote exchanges of Europeans and Americans with their homologues from Russia, China, Brazil, India, Asia, Australia, and Middle East… Being closer to communities and having a stronger impact at peripheral levels, these new transatlantic exchanges would be far more influential in the medium and longer term than national exchange programs should grow as a major action line for European policies and transatlantic cooperation.

Furthermore, *new technologies* should be the basis for innovative new working strategies. While the development of technologies and communication has more and more impact at social level and greatly influences younger generations, it is of crucial importance to provide content for 'technological art' that reflects the heritage of European and American cultural diversity and specifically addresses younger generations. Technologies should serve cultural understanding and online strategies should reflect the full range of possible contributions to cultural diplomacy on both sides of the Atlantic.

Finally, *innovative transatlantic alliances* should also take place in this agreement, giving space to experiment and explore—e.g., partnerships based on a revised pattern of 'mutuality'. In facts, 18 of the top 20 knowledge regions in the world are in the U.S. and in Europe. In terms of the Transatlantic innovation economy, as Daniel Hamilton highlighted,[23] 9 of the global top 20 companies spending on innovation are in the U.S. and 7 in Europe, when U.S.-based companies account for 34.3 % and EU-based companies for 30.6 % of the top R&D companies in the word with Japan accounting for an additional 22 %.

Conclusion

Although it has almost become a *cliché* to highlight the importance of culture in the conduct of modern diplomacy, more than ever culture has a major role to play, being both a diplomatic tool and "an indispensable bridge that leads diplomats into the hearts and minds of their respective audiences." Information and communication technology, global mass media outlets provide a solid ground for increased multinational dialogue and interaction, when traditionalism puts a cultural strain on such dialogue.

Despite many criticisms over the last 10 years, culture still has a critical role to play as a major forum for negotiation and a medium of exchange in finding shared solutions. Its impact on the conduct of modern diplomacy is unquestionable but should be reconsidered based on the current requirements of society and the everchanging world we live in. At a time where the EU and the U.S., despite the recession, remain each other's most important markets and face many common global challenges, they should establish a cultural and scientific transatlantic strategy and partnership. They could even be launched in the framework of the *Transatlantic Legislator's Dialogue*, between the European Parliament and the American Congress.

[23] Hamilton DS, Quinlan JP (2011) The Transatlantic Economy 2011: Annual Survey of Jobs, Trade and Investment between the United States and Europe. Washington DC: Center for Transatlantic Relations.

Contents

Abstract

Major societal challenges of a global nature include climate change, efficient energy supply, environmental sustainability, and health care. Science & Technology Policy (S&T) policy is an essential contributor to dealing with these challenges; moreover, international cooperation and collaboration in S&T is vital to tackling these issues, since no single nation or even region is able to respond adequately by itself. Within this context, this book addresses recent developments in transatlantic S&T cooperation between the EU and the U.S.

The EU–U.S. relationship dates back to the 1950s, with regular EU–U.S. Summits to assess and develop transatlantic cooperation. In the area of S&T, the EU and U.S. concluded an S&T Cooperation Agreement in 1998, renewed it in 2004, and extended it for another 5 years in July 2009. The scope of the cooperation has been enlarged, including a range of fields relating to security and space research. Each region has also pursued separate S&T strategies, as well. For example, in Europe, since the adoption of the Lisbon Strategy in the year 2000, the EU has committed itself to building a European Research Area (ERA) that extends the single European market to the world of S&T, ensuring open and transparent "trade" in S&T skills, ideas, and know-hows. It includes programs that enhance Europe's access to worldwide scientific expertise, attract top scientists to work in Europe, and contribute to international responses to shared problems. In the U.S., the Report, "Rising Above the Gathering Storm," the American Competitiveness Initiative, and the "Strategy for American Innovation" had a similar motivation, which was reinforced by the financial crisis. With differences in organizational structures, truly coordinated international S&T cooperation has not yet come to play a substantial role, though a variety of initiatives are underway to foster and promote collaboration.

The research underlying this study is based on interview with key stakeholders in the field, with an emphasis on:

- potential new opportunities and new mechanisms for increased transatlantic EU–U.S. S&T cooperation under current conditions
- examples of coordinated "science diplomacy" efforts
- options for the development of effective joint efforts.

While the project is focused on European–U.S. relationships, it also addresses issues of international S&T cooperation involving other regions, including Africa and Asia. The author highlights the urgency of S&T cooperation to address global issues, and the evolving roles of government, universities and research centers, and industry, in promoting successful strategies and programs.

Chapter 1
Introduction

Abstract The introduction to this report provides more general input why transatlantic science and technology cooperation is still relevant in today's globalized world. Today's major societal challenges are global in nature such as climate change, energy supply, efficient energy use, and global health issues. Science and technology (S&T) play a key role in finding solutions. Due to their global nature, these challenges demand collaborative approaches as any individual nation or region has neither the intellectual capacities nor the financial resources to respond adequately and effectively by itself. Thus, international cooperation in S&T has become more imperative than ever. Europe and the U.S., the two major mature economic powerhouses are in the position to take the initiative and assume responsibility to address mentioned global challenges via enhanced S&T cooperation. This report is based on a Fulbright-Schuman Scholarship implemented at two leading universities in Washington DC. Set up as a broad qualitative approach, the study systematically explored "real time" views, perceptions, and expectations of the U.S. and European key stakeholders in the field of strategic transatlantic S&T cooperation.

1.1 Why Bother: Is Transatlantic Science and Technology Cooperation Still Relevant in Today's Globalized World?

Today's major societal challenges are global in nature. Some of the most pressing challenges we face are related to climate change, energy supply, efficient energy use, and global health issues. Science and Technology (S&T) play a key role in finding solutions to all of these challenges. Due to their global nature, these challenges demand collaborative approaches as any individual nation or region has neither the intellectual capacities nor the financial resources to respond adequately and

S. E. Herlitschka, *Transatlantic Science and Technology*, SpringerBriefs in Business,
DOI: 10.1007/978-1-4614-4385-8_1,

effectively by itself. Thus, international cooperation in S&T has become more imperative than ever.

Who else, if not the two major mature economic powerhouses—Europe and the U.S.—should be the ones to take the initiative to address aforementioned global challenges via enhanced S&T cooperation? Representing approximately 54 % [1] of the world's GDP in 2011, their transatlantic partnership not only provides opportunities but should also assume responsibility for tackling these global challenges, as discussed repeatedly on the occasion of the recent EU–U.S. Summits.[2].

However, one has to ask the question if in today's globalized "S&T Enterprise," the transatlantic dimension—in particular the cooperation of Europe and the U.S.—does it even still play a major role? Just by looking at the growth rates and dynamics in emerging economies, would not it be somewhat outdated to think of Europe and the U.S. as the world's drivers in science, technology, and innovation in any way? Is Europe, together with the U.S., still able and/or willing to provide the leadership for driving the development of solutions to global challenges? Or have other dynamics at global level as well as in Europe and the U.S. themselves—factors such as the financial crises or the rapid development of Asia, and China in particular—become much more important?

This is the context of an investigation and analysis of the current state of the transatlantic European and the U.S. partnership in S&T in the light of today' global challenges as presented in the present study. The study seeks to provide insight into the current conditions and defining dimensions of transatlantic S&T cooperation, and offers policy recommendations on how to tap the full potential of a future transatlantic S&T partnership.

1.2 Background of This Report: Fulbright-Schuman Project and Motivation

Above mentioned and similar questions were the starting points for putting together a proposal: to conduct a pragmatic interview-based study on the views and perceptions of key stakeholders toward effective transatlantic S&T cooperation in the light of global challenges.

After some 20 years of professional life as practitioner in various sectors—from industry to academia, from research to research management and funding—it was the author's interest to combine professional experience in European and international S&T cooperation with deeper insight into transatlantic cooperation and getting new impulses, some "fresh air" for the brain, this research proposal was

[1] Hamilton DS, Quinlan JP (2011) The transatlantic economy 2011: Annual Survey of Jobs, Trade and Investment between the United States and Europe. Washington DC: Center for Transatlantic Relations.

[2] Overview EU-U.S. Summits: http://www.eurunion.org/eu/Table/EU-US-Summits/.

submitted to the Fulbright-Schuman program, was successfully evaluated and hence the grant was awarded.

Originally initiated by the U.S. Senator William Fulbright after World War II for the "promotion of international good will through the exchange of students in the fields of education, culture and science," the Fulbright Program has become the most widely recognized and prestigious international exchange program in the world, supported for more than half a century by the U.S. and its respective partners worldwide. The Fulbright-Schuman Program in particular is dedicated in advancing the understanding of the European Union (EU) and in promoting U.S.–EU relations. Financed by the U.S. Department of State and the European Commission Directorate-General for Education and Culture, it awards grants to academics and non-academic professionals of the U.S. and EU Member States.

This project—the study on the views and perceptions of key stakeholders toward effective transatlantic S&T cooperation in the light of global challenges—was implemented between Nov. 2010 and April 2011 at George Washington University/Elliot School of International Affaires and Johns Hopkins School of Advanced International Studies at the Center for Transatlantic Relations.

1.3 Objectives and Orientation of This Report

On purpose this report on "Transatlantic European–U.S. Science and Technology Cooperation" is presented in this book series "Innovation, Technology and Knowledge Management".

The aim of this book series is "to highlight emerging research and practice at the dynamic intersection of innovation, technology, and knowledge management, where individuals, organizations, industries, regions, and nations are harnessing creativity and invention to achieve and sustain growth."Volumes in this series explore "the impact of innovation at the "macro" (economies, markets), "meso" (industries, firms), and "micro" levels (teams, individuals), drawing from related disciplines such as finance, organizational psychology, research and development (R&D), science policy, information systems, and strategy, with the underlying theme that in order for innovation to be useful it must involve the sharing and application of knowledge."[3]

These are broader issues of discussion, to which this report contributes through its particular focus as pragmatic qualitative study built on first-hand input from key stakeholders in transatlantic S&T cooperation.

What makes this report unique…

The report is unique as it is based on a study that systematically explored "real time" views, perceptions, and expectations of the U.S. and European key

---

[3] Foreword by the Series Editor on the new book series Innovation, Technology and Knowledge Management, www.springer.com.

stakeholders in the field of transatlantic S&T cooperation. Set up as a broad qualitative approach the study has been built on interviews as well as extensive information via up-to-date reports, conferences and events.

Core questions included:

- What are the status, relevance of, and the experiences from former or current transatlantic S&T cooperation?
- Which opportunities and new mechanisms for increased and effective transatlantic European–U.S. S&T cooperation could be developed under the present conditions?
- What would be options for effective joint transatlantic efforts in the light of global challenges?
- What about coordinated transatlantic "science diplomacy" efforts?

The results of the study are summarized in this report, comprising the main findings as well as drawing conclusions and providing policy recommendations toward the European Commission and the contributors to this study. In particular funding organizations as well as Science Counselors have expressed substantial interest in the results of the study.

...but still a "snapshot" in the wider field of transatlantic S&T cooperation

As much as the results of this report are derived from broad first-hand evidence, it still represents a "snapshot" in a much wider context with a great variety of activities going on between Europe and the U.S. at all levels, from academia, business, NGOs to governments. Thus, this report is meant to contribute a piece to the puzzle of transatlantic S&T cooperation, providing insights at a time when science, technology, innovation, and knowledge management play a decisive role in the global knowledge economy.

Who is Europe...

In this study, "Europe" refers primarily to the EU and its more substantial, typically strategic activities in S&T at the EU level. Due to the focus on major global challenges, individual scientist-to-scientist cooperation was not taken into consideration for the study. As for the activities of single EU Member States, these were considered to some degree, although not in a comprehensive way.

Focus on strategic European–American cooperation...

A broad range of joint S&T activities between Europe and America is going on and has been developed over the years, most notably in two ways. On the one hand, scientist-to-scientist cooperation, measured by co-authorship in publications: European–U.S. cooperation is the strongest worldwide with more than 435,000 join publications between 2000–2009 and an annual growth rates of approx. 6 % over the last years.[4] One the other hand, generations of European students and postdocs have been a major factor for both sides—for Europe as its researchers

[4] European Commission (2011) Innovation Union Competitiveness Report Part 2, page 290.

took advantage of the training opportunities in the U.S., for the U.S. as its research enterprise largely depends on the foreign S&T workforce.[5]

This study does not address above mentioned scientist-to-scientist cooperation which takes place anyway and extensively. Instead, it looks at the transatlantic cooperation with respect to more substantial issues, typically bigger in dimension and impact due to their global relevance and as such of strategic nature. In short, this study explores European and U.S. views on issues where both sides can/may have a greater impact through more coordinated efforts.

While the study focused on strategic European–American relationships, it also addressed issues of international S&T cooperation in which both Europe and the U.S. are responding (or should/could respond) to joint opportunities and/or responsibilities in other parts of the world, e.g. Africa, Asia, etc.

Further impact…

The results of this study will also provide immediate impact to the strategic EU–U.S. project "BILAT-USA—Bilateral coordination for the enhancement and development of S&T partnerships between the European Union and the United States of America".[6] The "BILAT-USA" Project has been set up and funded by the European Commission in order to further strengthen the transatlantic dialog platform and set up a spectrum of pragmatic services to foster real S&T cooperation. This Fulbright study provides insights into underlying notions, reflections, and information contributing to a more effective implementation of activities such as "BILAT-USA".

1.4 Methodology Applied and Quantitative Indication

The methodological design of this study consists of a mix of several qualitative methods including expert interviews, questionnaires, and broad information gathering activities, thus ensuring and increasing the validity of the results.

The project was carried out in three stages:

As a first stage, some basic hypotheses were tested, and relevant questions on European–U.S. S&T cooperation were identified. A key element of the second stage was the survey with leading European and the U.S. stakeholders/experts in the field. As a third stage a small group of experts was invited to review the results.

---

[5] Based on National Science Board (2012) Science and Engineering Indicators 2012. Arlington VA: National Science Foundation (NSB 12-01).

[6] The BILAT-USA Project aims to improve the awareness towards EU–U.S. Science & Technology cooperation through setting up a sustainable, knowledge based, and bi-regional dialog platform between S&T key players as well as stakeholders from the EU-Member States, Associated Countries and the U.S. The Project is funded by the European Union 's Capacities Programme on International Cooperation under the seventh Framework Programme. More information: http://www.euussciencetechnology.eu/bilat-usa.

It is underlined that the results of this study are NOT based on a representative quantitative sample. Instead, key stakeholders were addressed and asked for their views. Similar to "focus group" approaches, the results provide information of trends and views supported by quantitative indications.

The questionnaires applied are included in Annexes 1 and 2.

A broad evidence base generated…

Overall, the results of this study have been derived from

- 80 extensive personal contacts to key stakeholders (interviews, questionnaires)
- More than 40 events/conferences/hearings related to the wider study issues
- More than 80 relevant published reports, papers and further references.

The most in-depth input resulted from systematic personal contacts via interviews and questionnaires, respectively. High profile interview partners were selected based on their personal or institutional expertise and relevance for the study topics. Intentionally, an approach was chosen to go beyond the "usual suspects" actors and to cover a broad range of institutions involved one way or the other in transatlantic S&T cooperation. Moreover, there has been more emphasis on input from the U.S. key stakeholders.

Organizations of contributors to this study are listed in Annex 3.

It has to be noted that contributors' input does not necessarily represent their respective, "official" institutional position, as the purpose of this study was not to produce an overview on official institutional positions. Since this report is not about individual opinions either, but instead about trends of notions, opinions and views, the names of individual contributors are not disclosed.

Chapter 2
Is European–U.S. Science and Technology Cooperation "Fit" to Working Jointly on "Grand Challenges"?

Abstract European–U.S. science and technology (S&T) cooperation takes place in an "ecosystem" with an abundance of so-called "grand challenges," typically major challenges of global dimension and in a setting of ever-increasing complexity at all levels. In this chapter, the wider context and major developments are described as they are important with their potential to affect, influence, or even shape transatlantic S&T cooperation, for instance, in the form of strategically joining S&T efforts, pooling resources, or developing new S&T-driven approaches to grand challenges issues. The wider context of transatlantic S&T cooperation is characterized by the economic crisis that is not just economics, innovation-related trends and their links to grand societal challenges, the competition for the best "brains," and the need to concentrate on core competencies in an increasingly networked world. In economic and several other dimensions, Europe and the U.S. are each others' most important partner; even though not commonly perceived as such, facts and figures provide an overview. Finally, the background and framework of European–U.S. S&T cooperation are summarized and described.

2.1 The Wider Context: "Grand Challenges" at All Levels

We do not "suffer" from a lack of "grand challenges," this seems to be a given fact of our times. On the contrary, there is an abundance of so-called "grand challenges", typically major challenges of global dimension—with even the wording becoming "inflationary" as everyone seems to talk about "grand challenges"—in a setting of ever-increasing complexity at all levels.

In this chapter, the wider context and major developments are described as perceived by the author. They are important as they have the potential to affect, influence, or even shape transatlantic S&T cooperation, for instance, in the form of

S. E. Herlitschka, *Transatlantic Science and Technology*, SpringerBriefs in Business,
DOI: 10.1007/978-1-4614-4385-8_2,

strategically joining S&T efforts, pooling resources, or developing new S&T-driven approaches to grand challenges issues.

The Economic Crisis: not just economics...

The financial crisis that started in the second half of 2008 was not only substantial in dimension but also had significantly broader impact, going way beyond economics. As Joe Quinlan, Fellow at the Transatlantic Academy put it recently "The 'Made in America' crisis also undermined the capacity and credibility of the world's economic architects—the United States and Europe, or the transatlantic partnership. After years of living beyond their means and after amassing mountains of debt, the music finally stopped for an economic alliance that had long set the tune for the global economy and grown accustomed to standing at the pinnacle of the global economic order."[1]

The most significant impact resulting from this financial crisis has been a developing substantial loss of credibility of so far perceived global economic governance mechanisms, projecting along this line, a likely loss of credibility of more general global governance mechanisms and models, summarized best as follows:

> The financial crisis accelerated a number of key long-range trends that were already in motion before the crisis struck. The relative economic decline of the developed nations and the rising influence of the emerging markets in general and China in particular were fast-forwarded by the crisis and have, in turn, accelerated the move toward a less U.S.-centric, more multi-polar world. While the global economy has recovered from the crisis, we are not going back to "business as usual." The new world before us will be more complex, fluid, and disruptive—notably for the architects of the post-war economic system. The United States and Europe have lost control of the global economic agenda and, critically, no longer control the key inputs of economic growth—labor, capital, and natural resources. These inputs are increasingly concentrated in the developing nations, who have emerged from the crisis more confident and emboldened. The future of globalization will be less U.S.-centric and more encompassing of Chinese, Turkish, Brazilian, and other characteristics of the developing nations. This phase of globalization heralds both promise and peril for the transatlantic partnership.[2]

Innovation, "Open" Innovation and grand societal challenges...

Innovation and societal challenges have become central points and a defining framework in almost every discussion on future orientations, competitiveness, or jobs.

In Europe, the course was newly set in 2010. The adoption of the "Europe 2020" strategy[3] conveys a clear signal as it is focused on "smart, sustainable and inclusive growth," linked to concrete objectives and goals. Seven so-called "Flagship Initiatives" are foreseen in order to implement the Europe 2020 Strategy.

---

[1] Quinlan J (2011) Losing control: the transatlantic partnership, the developing nations and the next phase of globalization. Transatlantic Academy.

[2] Quinlan J (2011) Losing control: the transatlantic partnership, the developing nations and the next phase of globalization. Transatlantic Academy.

[3] European Commission (2010) A strategy for smart, sustainable and inclusive growth. COM(2010) 2020.

One of them, the "Innovation Union"[4], lays out the strategic plan for research, technology, and innovation for the coming years. It represents a radical shift to innovation at all levels by concentrating on strengthening the innovation pipeline, with the overarching goal of improving the financial framework for research and innovation. More specifically, in the words of the responsible EU Commissioner for Research, Innovation, and Science: "the Innovation Union aims to do three things: First, transform Europe's world class science base into a world-beating one. Second, make coherent use of public sector intervention to stimulate the private sector. Third, remove the remaining bottlenecks to the commercialization of good ideas. The Innovation Union dedicates and entire chapter to boosting international cooperation, recognizing that working better with our international partners means opening access to our R&D programs, while ensuring comparable conditions abroad"[5]

Furthermore, innovation plays an important role in several of the other Flagship Initiatives, including the "Energy 2020" strategy,[6] the "Digital Agenda for Europe"[7]], and has been highly relevant in already existing activities such as the seventh Framework Program, and Joint Technology Initiatives combining EU and Member States efforts in areas such as innovative medicines, embedded computing systems, and nanoelectronics.

The new multiannual EU Framework Program for Research and Innovation "Horizon 2020" currently under preparation is a bold statement in "challenging" economic times. Proposed by the European Commission with a budget of €80 billion for the years 2014–2020 it has been set up to deliver jobs and strengthen competitiveness through funding for research and innovation. As such, Horizon 2020 complements the efforts of EU Member States to increase investment in research and innovation. More specifically, Horizon 2020 will provide investment to raise the level of excellence in Europe's science base, industrial technologies, and topics related to societal challenges. As its predecessor programs Horizon 2020 will be broadly open for international cooperation.[8]

In the U.S. innovation is high on the political agenda, also driven by importance and urgency. Or as Fareed Zakaria put it recently:

> Innovation is as American as apple pie. It seems to accord with so many elements of our national character—ingenuity, freedom, flexibility, the willingness to question conventional wisdom and defy authority. But politicians are pinning their hopes on innovation for more urgent reasons. America's future growth will have to come from new industries that create new products and processes. Older industries are under tremendous pressure. Technological

[4] European Commission (2010) Europe 2020 flagship initiative innovation union. COM(2010) 546 final.

[5] EU commissioner for research, innovation and science Maire Geoghegan-Quinn Speech on 18 January 2012, Washington DC.

[6] European Commission (2010) A strategy for competitive, sustainable and secure energy. COM (2010) 639 final.

[7] European Commission (2010) A digital agenda for Europe. COM (2010) 245 final.

[8] European Commission (2011) Horizon 2020: the framework programme for research and innovation. COM(2011) 808 final. http://ec.europa.eu/research/horizon2020/index_en.cfm.

change is making factories and offices far more efficient. The rise of low-wage manufacturing in China and low-wage services in India is moving jobs overseas. The only durable strength we have—the only one that can withstand these gale winds—is innovation.[9]

The U.S. President Obama himself is perceived highly interested in science, technology, and innovation, as reflected by a key stakeholder's statements in the course of the study presented here, as well as by the policies put forward by the Obama administration. The 2009 "Strategy for American Innovation"[10] as well as the 2011 update provides the direction and has been developed with a broad vision: "Our vision of America's future is one where prosperity is built by skilled, productive workers and sound investments that will spread opportunity at home and allow this nation to lead the world in the technologies, innovation and discoveries that will shape the 21st century."

It comprises three elements that cover a large part of what could be called the "innovation value chain," including investments in the building blocks of American innovation, the promotion of competitive markets that spur productive entrepreneurship, and catalyzing breakthroughs for national priorities.

The "America Competes Act" in its reauthorization version 1010 was signed by the U.S. President in January 2011. It is seen as a milestone in the path to building "an innovation economy for the 21st century—an economy that harnesses the scientific and technological ingenuity that has long been at the core of America's prosperity and applies that creative force to some of the biggest challenges we face today. Whether it's developing new products that will be manufactured in America, or getting and using energy more sustainably, or improving health care with better therapies and better use of information technology, or providing better protection for our troops abroad and our citizens at home, innovation will be key to our success."[11] As such, the America Competes Act specifies activities in line with the mentioned objectives.

In his State of the Union Address 2011, President Obama put a strong focus on innovation, underlined by the fact that the words "innovation"/"innovate" were mentioned 11 times, with one of the strongest statements as follows:

"We need to out-innovate, out-educate, and out-build the rest of the world…The first step in winning the future is encouraging American innovation. None of us can predict with certainty what the next big industry will be or where the new jobs will come from. Thirty years ago, we couldn't know that something called the Internet would lead to an economic revolution. What we can do—what America does better than anyone else—is spark the creativity and imagination of our people. We're the nation that put cars in driveways and computers in offices; the nation of Edison and

[9] Zakaria F (2011) The Future of innovation: can America keep pace? Time U.S.:http://www.time.com/time/nation/article/0,8599,2075226,00.html.

[10] A strategy for american innovation: driving towards sustainable growth and quality jobs (2011) http://www.whitehouse.gov/innovation/strategy.

[11] America competes Act, 2010, http://www.whitehouse.gov/blog/2011/01/06/america-competes-act-keeps-americas-leadership-target.

the Wright brothers; of Google and Facebook. In America, innovation doesn't just change our lives. It is how we make our living."[12]

The concept of "Open Innovation"—which has become a buzzword in recent years—summarizes another major development: innovation processes representing a complex interaction and exchange of various actors, including companies, academia, markets, and users. Initially published by Henry Chesbrough in 2003, and based on his research into the innovation practices of large multinational companies, "Open Innovation" describes a new paradigm for the management of industrial innovation in the twenty-first century. According to this paradigm, firms work with external partners to commercialize their internal innovations and to obtain a source of external innovations that can be commercialized.[13]

The paradigm of "Open Innovation" has influenced almost every sector. "Open Innovation" has become a symbol for active exchange and interaction with the entire surrounding field and its respective "markets." Most of today's strategy reflections rely on "Open Innovation" implicitly or explicitly. "Open Innovation" is discussed in various forms in education[14,15] as much as with regard to universities, funding agencies, etc., even though it is not necessarily new in every field as can be demonstrated by the role of the "innovation benefactor".[16] "Open Innovation" even has become an issue in very different cultures and settings such as China.[17,18]

Taking it all together, the bottom line of current trends is that while the aim of basic research still is to discover new knowledge and as such plays an essential role, applied science and technology seem to become more oriented toward innovation. Science, technology, and innovation themselves are more and more understood as having to deliver answers to societal questions and challenges.

People and diversity, the real "battlefield"…

What really matters is always people, particularly in dealing with creativity and inspiration as required and found in education, science, technology, and innovation. The competition for the best brains has become a global one; expressions like "brain drain, brain gain, brain circulation" have become common and reflect what has become the name of the game.

---

[12] U.S. President Obama's State of the Union Address 25 January 2011: http://www.whitehouse.gov/the-press-office/2011/01/25/remarks-president-state-union-address.

[13] Open Innovation Community: http://www.openinnovation.net.

[14] Oliveira Santos C (2011) Open innovation in education: the open design innovation project: http://library.iated.org/view/OLIVEIRASANTOS2011OPE.

[15] Flagship Initiative, Collaboration: Open Innovation Web Portal: http://www.ed.gov/open/plan/flagship-initiative-collaboration.

[16] Chesbrough H (2003) The era of open innovation. Sloan Management Review, 44(3): 35-41

[17] Liu X, Lundin N (2006) Toward a market-based open innovation system of china: http://www.globelicsacademy.net/2007/papers/Xielin%20Liu%20Paper%201.pdf.

[18] Fu X, Xiong H, (2011) Open innovation in China: policies and practices. Journal of Science and Technology Policy in China, Vol. 2 Iss: 3, pp.196–218.

Virtually every organization, region, and nation is reaching out for the "best and brightest" and does so with specific initiatives and programs, several of them devoted particularly to their national "diaspora" abroad. Cultural "imprinting" seems to play an important role and obviously differs in comparison to industrialized countries and Asia, of whom a bigger share of their researchers seems to be highly inclined to return home after years of training/working abroad. Particularly, China is more aggressively pursuing this strategy, because these are the people who represent the truly limiting factor in striving for sustainable competitiveness.[19,20,21,22]

The issue of the "best and brightest" is directly linked to diversity and its appreciation. Diversity understood as differences in gender, age, social context, culture, ethnics, or religion will have to be more actively embraced as a key opportunity in terms of societal values as well as the success factor for competitiveness. Society as a whole—even in the short run—will just no longer be able to afford not taking advantage of diversity. This is particularly true for the role of women in S&T in many countries.

Western, industrialized societies seem to "loose" their young generation.[23] Education and knowledge orientation seem to loose in relevance, particularly in STEM disciplines (Science, Technology, Education, and Mathematics). Both in Europe as well as in the U.S. massive initiatives to tackle this development have been put forward and are on the way of being implemented at all levels. However, there seems to be a huge difference in the appreciation for education, for instance, in Asia as compared with industrialized countries, strikingly as well as controversially described in the "Tiger Moms",[24] but evidently obvious for everyone traveling through Asia.

Particularly, in the light of societal as well as demographic developments, countries will improve their chances for future competitiveness that are capable to offer

- Attractive opportunities for the "best and brightest,"
- Societies welcoming intercultural exchange,
- Proactive management capabilities at all levels for diversity in its various appearances.

---

[19] Cao C, Suttmeier R (2001) China's New Scientific Elite: Distinguished Young Scientists, the Research Environment and Hopes for Chinese Science. The China Quarterly.

[20] Xiaoming L (2011) Rise in scientists returning to China. Nature 475, 296. doi: 10. 1038/475296d.

[21] LaFraniere S (2010) Fighting trend, China is luring scientists home. New York Times. http://www.nytimes.com/2010/01/07/world/asia/07scholar.html?pagewanted=all.

[22] Wadhwa V, Jain S, Saxenian A, Gereffi G, Wang H (2011) The grass is indeed greener in India and China for returnee entrepreneurs. Ewing Marion Kauffman Foundation.

[23] European Commission (2010) Youth on the move: an initiative to unleash the potential of young people to achieve smart, sustainable and inclusive growth in the European Union. COM(2010) 477 final.

[24] Chua A (2011) Battle Hymn of the Tiger Mother. The Penguin Press.

While already today leading knowledge-based organizations have trouble finding highly qualified people, this trend will intensify substantially and globally with the demographic and societal (for instance on how to deal with migration) developments ahead of us.[25]

Core competencies in a networked world…

"Networking" is another keyword of our time. Globalization in the sense of interconnectedness and cooperation across regions, nations, sectors, disciplines, and organizations has become the daily practice, even more so in the "knowledge arena" of science, technology, innovation, and education. For those who are capable of doing so, it has become extremely easy to get connected to everyone around the world.

To use Thomas Friedman's words: "In the Cold War, the most frequently asked question was: 'Whose side are you on?' In globalization, the most frequently asked question is: 'To what extent are you connected to everyone?'"[26]

Excellence in its various expressions frequently develops in crossing disciplines and sectors, and thus is significantly influenced by networking. At the same time, interdisciplinarity, cooperation of companies and academia are still issues of concern, at least attention, and still prove to be challenging.

As much as the level of interconnectedness increases, the trend to concentrate on core competences is also reinforced. Organizations, regions, and nations think about their strengths as well as weaknesses, and embark on comprehensive strategy development processes in order to invest limited resources in a more targeted way, strengthening core competences, and thus trying to increase their competitiveness. Likewise, this is true for companies, universities, and research organizations.

One of the most important questions in this setting of vast global opportunities is "who are the right partners to cooperate with," that are matching best their own core competences and hence contributing to mutual benefits.

Intercultural cooperation is an essential component and a decisive competitive advantage; it is the key to unlocking and effectively using this capability in developing cooperation across sectors, disciplines, nations, genders, and cultures. Intercultural cooperation itself is driven by factors such as globalization and the increasing complexity and speed of interactions in all areas, from economics to society to technology, to name only a few.

The bottom line in a globalized world—particularly in science, technology, and innovation—with ever-increasing opportunities and challenges is the ability to effectively build, manage, and expand a network of key partners and to identify the most suitable partners based on their expertise. This has become THE competitive advantage.

---

[25] The Economist (2006) The battle for brainpower.

[26] Friedman T (1999) The lexus and the olive tree, understanding globalization. Farrar, Straus and Giroux.

2.2 Europe and the U.S.: Each Others' Most Important Partner, Yet

A number of highly relevant reports on the transatlantic cooperation, its potentials and requirements have been published recently. As the general "ecosystem" of transatlantic relations is considered an important framework condition for S&T, key facts, figures, and trends are presented here, not only in essence but through strong convincing quotes summarized in this chapter.

The Transatlantic Economy 2011[27] analysis provides a comprehensive overview on the status of the transatlantic cooperation as follows:

"The United States and Europe remain each other's most important foreign commercial markets, a fact still not fully appreciated by opinion leaders on both sides of the transatlantic. Put simply, no other commercial artery in the world is as integrated and fused together as the transatlantic economy. Ties are particular thick in foreign direct investment, portfolio investment, banking claims; trade in goods and services, onshore jobs, and flows of ideas in terms of mutual R&D investment; patent cooperation; technology flows; and sales of knowledge- intensive services. These deep economic bonds have been critical to prosperity on both sides of the Atlantic. Yet as both the EU and the United States struggle to recover from the economic and financial crisis, such ties can also amplify the challenges of slow growth."[28]

Europe and the U.S. in economic terms: some facts are summarized in Box 2.1.

Box 2.1: Facts and Figures on European–U.S. Cooperation in Economic Terms, based on[29,30]

- Despite the recession, the U.S. and Europe remain each other's most important foreign commercial markets. No other commercial artery in the world is as integrated and fused together as the transatlantic economy
- The transatlantic economy is the largest and wealthiest market in the world, accounting for over 54 % of world GDP in terms of value and 40 % in terms of purchasing power.

---

[27] Hamilton DS, Quinlan JP (2011) The transatlantic economy 2011: annual survey of jobs, trade and investment between the united states and europe. Washington DC: Center for Transatlantic Relations.

[28] Hamilton DS, Quinlan JP (2011) The transatlantic economy 2011: annual survey of jobs, trade and investment between the United States and Europe. Washington DC: Center for Transatlantic Relations.

[29] Hamilton DS, Quinlan JP (2011) The Transatlantic Economy 2011: Annual Survey of Jobs, Trade and Investment between the United States and Europe. Washington DC: Center for Transatlantic Relations.

[30] Hamilton DS (2011) Europe 2020, Competitive or Complacent? Center for Transatlantic Relations, Washington DC.

- Ties are particularly thick in Foreign Direct Investment, portfolio investment, banking claims, trade, and affiliate sales in goods and services, mutual R&D investment, patent cooperation, technology flows, and sales of knowledge-intensive services.
- North America is the largest regional destination of EU Foreign Direct Investment (FDI) and the largest regional source of FDI in the EU. The EU is the top destination of the U.S. FDI around the world. The U.S. FDI in the EU of over €1 trillion is more than the next 20 investors combined.
- The EU FDI of €1.25 trillion in North America is more than the next six destinations combined.
- The U.S. and Europe are the two leading services economies in the world. The U.S. is the largest single country trader in services, while the EU is the largest trader in services among all the world regions.
- Of the global top 20 companies spending on innovation, nine are in the U.S. and seven in Europe.
- 18 of the top 20 knowledge regions in the world are in the U.S. and Europe.

"However, the world that created the transatlantic partnership is fading fast".[31] "If the U.S.-European relationship is to be a progressive force in the world to come, Americans and Europeans must urgently build a more strategic partnership that is more effective in dealing with new partners and addressing transformations occurring all around them. It is a moment of opportunity—to use or to lose."[32]

Recent discussions on prospects for cooperation in the Atlantic Basin are promising:

"Meanwhile Northern Atlantic basin states are the architects of the post-war economic and security order—a liberal order whose foundational ideas remain more widely accepted today than its institutional architecture, which represents a snapshot of the distribution of power in 1945. In the absence of global agreement on reframing institutions of governance, it seems doubly important to examine the Atlantic space as a region ripe for better mechanisms of cooperation….Energy, climate change, and natural resources are a key theme in the Atlantic. The divergence between the most and least efficient producers (and the most and least prolific consumers) is perhaps greater than anywhere else on the planet. The North Atlantic states have technological solutions that are the most advanced in the world. Yet they cannot translate into control of agendas and solutions, or preservation of historical rights and access to common resources. Governance

---

[31] Hamilton DS, Barry C, Binnendijk H, Flanagan S, Smith J, Townsend J (2009) Alliance Reborn: An Atlantic Compact for the 21st Century, The Washington NATO Project, http://transatlantic.sais-jhu.edu/bin/i/y/nato_report_final.pdf.

[32] Hamilton DS (ed) (2010) Shoulder to Shoulder, Forging a Strategic U.S.-EU Partnership. Johns Hopkins University Center for Transatlantic Relations, Washington DC.

mechanisms for common resources have been devised in the North Atlantic. How can these be translated successfully to other parts of the basin?"[33]

"Nothing is more imperative for EU foreign policy than defining a common agenda with the U.S. Regrettably, in recent times transatlantic relations have all too often been marred by ideological divergences that are largely a legacy of the Cold War era. Such dissensions are clearly dysfunctional in today's multipolar world, which calls for a concerted effort to build broader coalitions that transcend ideological divides."[34]

The U.S. economic situation, the massive financial dependence on China, together with the geopolitical judgment has clearly shifted the U.S. perspective during the current Administration to Asia and the U.S. President calling himself the "First Pacific President."[35]

On Europe, various comments in this context as expressed in the course of this study can be summarized by: Europe is not part of the problem anymore, but it is not part of the solution either.

In essence, one conclusion can also be that the relationship of Europe and the U.S.—in recognition of the economic interconnectedness and potentials–needs the transition from an alliance originally after World War II set up AGAINST a joint enemy, toward a new identity and alliance FOR the development of joint solutions.

2.3 Background and Framework of European–U.S. Science and Technology Cooperation

This overview is focussed on European activities as represented by the European Union (EU). Although the EU–U.S. relationship dates back to the 1950s, the first formal framework for cooperation was put in place in 1990 with the Transatlantic Declaration, followed by a new Transatlantic Agenda,[36] and regular EU–U.S. Summits to assess and develop transatlantic cooperation.[37]

In the area of S&T, the EU and U.S. concluded an S&T Cooperation Agreement in 1998, renewed it in 2004, and extended it for another 5 years in July 2009. The scope of the cooperation has been enlarged, including security and space research among a range of other fields. However, despite political goodwill on both sides, in the EU and the U.S., it is interesting to see some opposite developments as well,

[33] European Commission (2011) The Atlantic Geopolitical Space: common opportunities and challenges.

[34] Kodmani B, Larrabee FS, Lewis P, Pawlak P, Rodrigues MJ, Vasconcelos A de (ed) (2011) The Agenda for the EU-U.S. strategic partnership. The European Union Institute for Security Studies, Paris.

[35] Asia Rise Drives Obama Message as U.S.'s First Pacific President, Businessweek, 11 Nov. 2011.

[36] http://www.eurunion.org/partner/agenda.htm

[37] Overview EU-U.S. Summits: http://www.eurunion.org/eu/Table/EU-US-Summits/.

particularly with respect to the general orientation toward real international S&T cooperation, and diverging focus on specific thematic areas such as climate change or security. The Economic Crisis and the broader political environment have contributed to more inward-looking attitudes both in the U.S. and in Europe, thus challenging also the transatlantic partnership.

In Europe, since the adoption of the Lisbon Strategy in 2000, the EU has committed itself to building a European Research Area (ERA) that extends the single European market to the world of S&T—ensuring open and transparent "trade" in S&T skills, ideas, and know how. As globalization at all levels demands that European research look outward, international S&T cooperation forms an integral part of EU S&T policy. It includes programs that enhance Europe's access to worldwide scientific expertise, attract top scientists to work in Europe, and contribute to international responses to shared problems. This openness to international S&T cooperation is also reflected in the Europe 2020 Strategy and related flagship initiatives agreed upon in 2010.

At an operational level, the EU research flagship program—currently the seventh Framework Program for Research, Technological Development and Demonstration (FP) for the years 2007–2013—has been expanded in scope and opportunities for international S&T cooperation, in addition to the massive funding made available under previous FPs. With its broad international perspective, the FP could, in principle, be understood also as a huge "science diplomacy" agent, opening and funding cooperation with all regions of the world. The structure of the successor program, the EU Framework Program for Research and Innovation—"Horizon 2020"—was proposed by the European Commission in 2011. Horizon 2020 is focused on further strengthening Europe's science base and research infrastructure, expanding Europe's leadership in key industrial technologies, and facilitating research and innovation that address societal challenges. With a proposed budget of €80 billion, foreseeing a 40 % budget increase despite economic "challenges," Horizon 2020 will again be open for international cooperation. Following discussions and negotiations, Horizon 2020 will be available for investing in research and innovation projects for the years 2014–2020.[38] On the occasion of her recent visit to Washington, D.C., the EU Commissioner for Research, Innovation, and Science stressed on the opportunities to strengthen transatlantic cooperation through Horizon 2020.[39]

In the U.S., still the world leader in S&T, the reports—despite their differing organizational settings and backgrounds–"Rising Above the Gathering Storm," the American Competitiveness Initiative, and the "Strategy for American Innovation" had a motivation similar to that of the Lisbon and Europe 2020 strategies. This motivation was reinforced by the financial crisis and major follow-up

---

[38] European Commission (2011) Horizon 2020 - The Framework Programme for Research and Innovation. COM(2011) 808 final. http://ec.europa.eu/research/horizon2020/index_en.cfm.

[39] EU Commissioner for Research, Innovation and Science Maire Geoghegan-Quinn Speech on 18 January 2012, Washington DC.

documents[40] to those mentioned above, as well as by statements including the State of the Union Address 2011 of President Obama where he stressed "the first step in winning the future is encouraging American Innovation"[41] and "we need to out-innovate, out-educate, and out-build the rest of the world."[42]

As much as the U.S. S&T system is organized differently and in a decentralized way, truly coordinated international S&T cooperation has not yet come to play a substantial role. In addition, security concerns plus related regulations seemed to have played a role in this respect. But there are promising developments as well. One example is the use of S&T cooperation in the context of science diplomacy as an effective smart power instrument. Moreover, initiatives such as the U.S.–EU Energy Council established in 2009 also indicate the intention to take new routes in strategic energy S&T cooperation. However, as similar approaches have been put in place with other major actors including China and India, the U.S.–EU Energy Council is expected to prove its added value and comparative advantages.

[40] Members of the 2005 "Rising Above the Gathering Storm" Committee (2010) Rising Above the Gathering Storm, Revisited: Rapidly Approaching Category 5. Prepared for the Presidents of the National Academy of Sciences, National Academy of Engineering, and Institute of Medicine: http://www.nap.edu/catalog.php?record_id=12999; America Competes Reauthorization Act, 2010: http://www.gpo.gov/fdsys/pkg/BILLS-111hr5116eas/pdf/BILLS-111hr5116eas.pdf, PCAST Reports: http://www.whitehouse.gov/administration/eop/ostp/pcast/docsreports.

[41] http://www.whitehouse.gov/the-press-office/2011/01/25/remarks-president-state-union-address.

[42] http://www.whitehouse.gov/the-press-office/2011/01/25/remarks-president-state-union-address.

Chapter 3
European–U.S. Science and Technology Cooperation Under the Current Conditions: The Study on Views, Reflections, Expectations of Key Stakeholders

Abstract In the light of today's globalized world—as described above—what do key stakeholders think of the status, potential, and opportunities of transatlantic European–U.S. S&T cooperation? This chapter provides a summary of key stakeholders views, perceptions, and expectations. Issues analyzed refer to the role and opportunities of European–U.S. S&T cooperation, experiences made so far, and the role of science diplomacy. In the light of today's globalized world—as described above—what do key stakeholders think of the status, potential, and opportunities of transatlantic European–U.S. S&T cooperation?

3.1 Setting the Scene: Identifying the Issues

As a first step a "draft picture" of stakeholders' views on key issues regarding transatlantic European–U.S. S&T cooperation was generated based on the stage 1 questionnaire (see Annex 1). This stage 1 questionnaire covered the following questions:

- What are key issues for transatlantic European–U.S. S&T cooperation: What are the most important questions regarding European–U.S. Science and Technology Cooperation.
- What are stakeholders' perceptions of promoting and limiting factors for European–U.S. S&T cooperation
- What are the stakeholders' perceptions on Europe's and U.S. openness toward international cooperation. Are Europe and the U.S. open for coordinated, more strategic international cooperation?

The findings of the stage 1 questionnaire are summarized as follows.

S. E. Herlitschka, *Transatlantic Science and Technology*, SpringerBriefs in Business,
DOI: 10.1007/978-1-4614-4385-8_3,

3.1.1 Key Issues for Transatlantic European–U.S. S&T Cooperation: What are the Most Important Questions Regarding European–U.S. Science and Technology Cooperation

Major issues mentioned by key stakeholders are summarized and categorized—following more overarching issues—according to levels of action encompassing governance, scientists', economic, global, and operational level.

Overarching issues relevant for transatlantic European–U.S. S&T cooperation include:

- The need to convince the U.S. political and economic leadership as well as scientific mainstream that transatlantic European–U.S. cooperation is truly in the U.S. interest. In doing so it is important to go beyond stereotypes about Europeans (such as "focussed on processes and institutional peculiarities often impenetrable for people", "inflexible labour markets", "underperforming economy", "mediocre universities and people who either strike or go on vacation") and Americans ("simplistic", "interested only in short-term business profits", "not understanding the complexity of systems", etc.).
- Consequential streamlining within Europe, leading to more visibility of European strength representing a region of 500 Mio. people through
  - Balancing of cooperation with respect to European Institutions and the EU Member States.
  - Centralizing research funding, thus creating counterparts to major U.S. Agencies/Funding organizations,
  - Real harmonization of European University System beyond the "Bologna Process",
  - Enabling the free flow of research money in Europe, thus creating a real European research "market place" of 500 Mio. people.
  - The consequential challenge for the U.S. would be to really accept Europe, treat it like a real partner, thus truly coordinate research efforts and avoid redundancies and potentially streamline funding.

Governance level:

- Key stakeholders were asking for the real benefit of European–U.S. Science and Technology Cooperation, as well as how a system can be created that facilitates S&T cooperation across the Atlantic and is in the interest of both, science and the public.
- Clear goals and concepts together with concrete overview on joint activities for cooperation based on real partnership and trust on both sides should be defined, along with functioning governance and funding structures.
- Focus on what is really important, the definition of clear goals and results. There is an inclination to think the process is more important than the outcome of real S&T results.

- Defining specific agendas by actively including the scientific communities.
- Defining, further developing mechanisms for cooperation that are effective and meet the needs of researchers facilitating transatlantic S&T cooperation.
- How to balance a more bottom-up S&T System in the U.S. with a more top-down approach as applied in the EU Framework Program?
- Cooperation in the fields of security and potentially in coordination with NATO.

Scientists' level:

- Cooperation in social sciences, the role of science in society, its wider impact, etc. has a lot of potential and should be implemented.
- Cooperation in the fields of large research infrastructures.
- Fostering the further mobility and exchange between scientists.

Economic level:

- Worldwide research & development expenditures are estimated $1,276 billion (purchasing power parities) in 2009, out of which 56 % are contributed by the U.S. and Europe.[1] Cooperation has enormous potential, particularly as a consequence of the financial crisis.
- EU–U.S. relations are trade focussed; S&T could contribute and has to be more actively considered.
- Innovation-related cooperation should be further developed based on strengthened S&T cooperation and mutual recognition of standards and regulations.

Global level:

- Traditional roles are changing: the U.S. perceives itself as a leader, thus is cautious in collaborations. These roles are changing in the light of global developments. What are the implications on Europe and the U.S.?
- Global challenges need global answers. However, all actors have to be included.
- Using Science Diplomacy in coordinated way would be important.

Operational level:

- Further increasing S&T cooperation through facilitating the framework for access and mobility of researchers.
- Develop more suitable, tailor-made tools for providing information, communication, and exchange.

---

[1] National science board (2012) Science and engineering indicators 2012. Arlington VA, National science foundation (NSB 12-01).

3.1.2 Perceptions of Promoting and Limiting Factors for European–U.S. S&T Cooperation

Key stakeholders' perceptions on promoting and limiting factors for transatlantic European–U.S. S&T cooperation are consistent to a large extent, as there are common views on the majority of issues concerning Europe as well as the U.S.

The common views and perceptions as regards Europe and the U.S. are summarized in the following Table 3.1 categorized according to levels of action encompassing governance, scientists, economic, global and operational level.

Differing views and perceptions on Europe and the U.S. were mentioned basically with respect to limiting factors for European–U.S. Cooperation. The points mentioned refer to internal structural aspects in the respective system in Europe and in the U.S.

As for the U.S., limiting factors were seen predominantly at governance level, particularly as regards:

- Legal (for instance ITAR-International Traffic in Arms Regulations) and visa issues,
- U.S. low level of international orientation,
- U.S. losing attractiveness for the "best and brightest",
- U.S. feels strong as leader and is reluctant to cooperate,
- Budgetary limitations due to the crisis.

As for Europe, limiting factors were seen primarily at governance as well as operational level, specifically as regards:

At governance level:

- Lack of European leadership to drive the agenda,
- European structures perceived as too complex and difficult,
- EU–Member States fragmentation.

At operational level:

- European–U.S. differing views on the role of science in society, differences of structures and focus areas.

3.2 Diving into the Issues: The Substance of the Study

The following, second stage of this study was implemented through a systematic survey among key stakeholders/experts in the field of transatlantic S&T cooperation.

It is underlined that the results of this study are not based on a representative quantitative sample. Instead, key stakeholders were invited to provide their views, perceptions, and experiences, thus contributing to a qualitative study. Similar to

Table 3.1 Key stakeholders' common perceptions on promoting as well as limiting factors for transatlantic European–U.S. S&T cooperation

Common views and perceptions on factors promoting or limiting European–U.S. Cooperation:	
Promoting factors	**Limiting factors**
At Governance level: Common vision, clear goals, and ideas of expectations of cooperation and governance Common interests, cultural proximity, economic linkages Mutual recognition of partners' capabilities, and understanding of mutual benefits Funding for cooperation	At Governance level: Lack of real joint funding, discussions on opening up of research funding programs have not resulted in concrete outcomes (with some exceptions such as health research) Underlying notions and differing drivers such as competition-orientation versus cooperation-orientation, timing, etc. Legal/contractual issues with respect to the U.S. participation in EU-Framework Program Misalignments of agency policies and funding cycles in Europe and the U.S. No adequate political platform for policy making and research programs despite existing Joint Consultative Meetings.
Scientists' level: People's personal contacts and mobility Bottom-up research activities	Scientists' level No direct contacts between researchers
Economic level: Economic conditions in times of budgetary limitations fostering pooling of resources (as opposed to budgetary limitations hindering cooperation) Discussions in the context of the Transatlantic Economic Council and Transatlantic Business Dialog	
Global level: Joining forces in dealing with global challenges: in coordinating major international cooperation efforts and in applying science diplomacy Global developments toward emerging economies	Global level: Country specific interests with respect to "opportunities" in emerging economies
Operational level: Mechanisms for cooperation in place, EU-Framework Program generally open to the U.S. researchers Shared data access	Operational level: Mechanisms, bureaucracy: not effective enough

"focus group" approaches, the results provide information on trends and views supported by quantitative indications.

A comprehensive questionnaire (see Annex 2) was used as basis for personal interviews which covered the following three main areas and related questions:

- Role and opportunities of European–U.S. S&T cooperation:
  - What role and effects are seen for European–U.S. S&T cooperation in the categories of advancing S&T, economic effects and competitiveness, international development, and sustainable development?
  - What kind of activities are considered useful and needed in order to improve the positive effects of European–U.S. S&T cooperation?
  - What role is seen for S&T education and mobility?
  - The current situation in Europe and the U.S. in S&T, political, and economic terms toward transatlantic as well as international S&T cooperation?
  - What effects did the security-driven considerations in the U.S. have on S&T cooperation?
  - Is the EU-Research Framework Program considered an effective approach toward S&T contributions to tackling major challenges and increasing competitiveness? Is it considered an effective approach to foster transatlantic S&T cooperation?
  - Thematic S&T fields for cooperative transatlantic activities?
  - Opportunities for transatlantic S&T cooperation
- Learning from Experiences:
  - What mechanisms of transatlantic S&T cooperation worked well, did not work well? What are useful examples as well as lessons learned?
- Science Diplomacy and European–U.S. cooperation:
  - Is there a need for joint activities?
  - Are there known examples?
  - What would be necessary in order to develop coordinated transatlantic Science Diplomacy activities?

The results based on the comprehensive second stage questionnaire are summarized as follows according to the three areas mentioned above.

3.2.1 Role and Opportunities of European–U.S. S&T Cooperation

S&T plays an ever increasing role in today's changing world; with emerging economies, global challenges, global networking, etc., Europe and the U.S. jointly still represent a partnership of approximately 54 %[2] of world GDP which not only provides great opportunities but also creates responsibilities for action.

[2] Hamilton DS, Quinlan JP (2011) The transatlantic economy 2011: annual Survey of jobs, trade and investment between the united states and europe. Washington DC, Center for Transatlantic Relations.

Question: What role and effects are seen for European–U.S. S&T cooperation in the following four specific categories: advancing S&T, economic effects and competitiveness, international development, and sustainable development?

Respondents provided comments on the following aspects, andresults are summarized in Table 3.2:

- How strong are the current effects of European–U.S. S&T cooperation seen with respect to above-mentioned categories of advancing S&T, economic effects and competitiveness, international development, and sustainable development?
- In which areas is a need seen for improvement of S&T effects including respective activities and examples?
- What kind of activities are considered useful and needed in order to improve the positive effects of European–U.S. S&T cooperation?
- What role is seen for S&T education and mobility?

The strongest effects of current European–U.S. S&T cooperation is seen in the category "advancing S&T" with the highest shares of substantial effects, with U.S. respondents being very clear, whereas European respondents assume effects equally distributed between "slight" and "substantial" effects. Nevertheless, 56 % of all respondents think that substantial improvements are needed in order to further advance S&T.

Slight effects of current European–U.S. S&T cooperation are seen in the categories "economics, competitiveness, innovation" as well as in "sustainable development". While 45 % of all respondents think that substantial improvements are needed in "economics, competitiveness, innovation", 80 % do so with respect to "sustainable development".

Although the category of "international development" is seen as least affected by current European–U.S. S&T cooperation, 65 % of key stakeholders think that substantial improvements are needed.

Table 3.2 summarizes the major results and comprises examples and activities perceived necessary in the respective categories, supplemented by aspects of S&T education and mobility.

Question: How could the current situation in Europe and the U.S. be described in S&T, political, and economic terms toward transatlantic as well as international S&T cooperation?

The majority of respondents — 42 and 46 %, respectively—thinks that transatlantic S&T cooperation takes place under favorable S&T conditions both in Europe as well as in the U.S. As regards international S&T cooperation 45 % consider conditions favorable in S&T terms in Europe and in the U.S.

The categories favorable in "economic" and "political" terms were similarly lower ranked, results are summarized in Fig. 3.1, and specified with respect to

Table 3.2 Role and effects seen for European–U.S. S&T cooperation in the following four specific categories: advancing S&T, economic effects and competitiveness, international development, and sustainable development

	Advancing science and technology	Economic effects and competitiveness including innovation	International development *(e.g. in dealing with developementissues in Africa, Asia, etc.)*	Sustainable development *(e.g., environment, energy, peace, etc.)*
How strong are the current effects of European–U.S. S&T cooperation seen with respect to categories of advancing S&T, economic effects and competitiveness, international development, and sustainable development? Figures indicate shares within their respective block in %. Categories of Columns: First column: no effects second column: slight effects third column: substantial effects	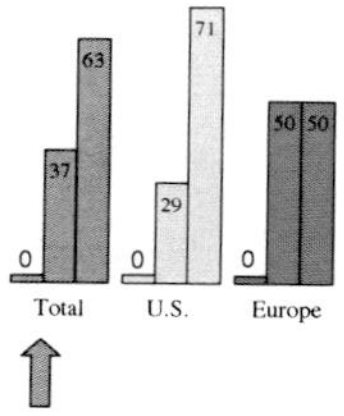 Total: 0, 37, 63 U.S.: 0, 29, 71 Europe: 0, 50, 50	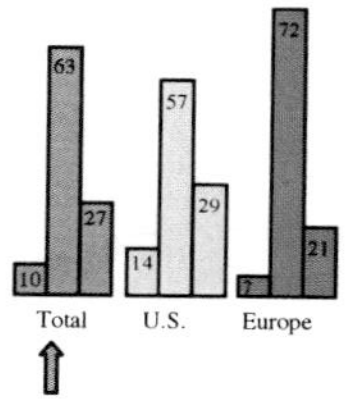 Total: 10, 63, 27 U.S.: 14, 57, 29 Europe: 7, 72, 21	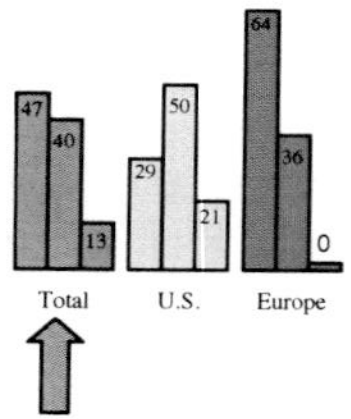 Total: 47, 40, 13 U.S.: 29, 50, 21 Europe: 64, 36, 0	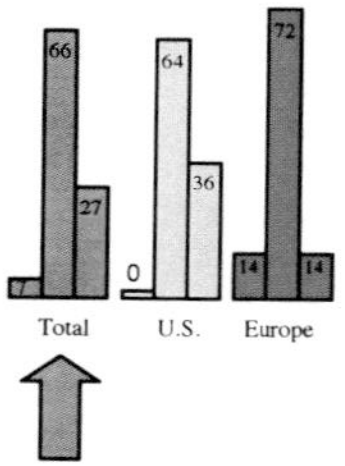 Total: 7, 66, 27 U.S.: 0, 64, 36 Europe: 14, 72, 14
In which areas indicated in this line is a need seen for improvement of S&T effects and thus respective activities? Categories: Substantial improvements Slight improvements The size of the arrows is positively correlated with the percentage.	56 % of all respondents think that substantial improvements are needed 38 % of all respondents think that slight improvements are needed	45 % of all respondents think that substantial improvements are needed 45 % of all respondents think that slight improvements are needed	65 % of all respondents think that substantial improvements are needed 31 % of all respondents think that slight improvements are needed	80 % of all respondents think that substantial improvements are needed 16 % of all respondents think that slight improvements are needed
What could serve as examples?	The level of bottom-up scientist to scientist cooperation between Europe and the U.S. is substantial and growing, positive effects also result from mutual competition. Spillover effects could be created toward competitiveness, international cooperation, and sustainable development. Space science could serve as example, as the more European and U.S. Space Agencies cooperate, the more optimal they work together (e.g., Mars exploration is being undertaken with NASA and ESA, built on successful international cooperative space science missions of the past).	There are so many barriers to economic cooperation in innovation that it is difficult to see how they can be overcome. Indirectly, a lack of cooperation might lead to a higher level of competitiveness for global leadership in economic and innovation issues. In the U.S.: a lack of recognition that commerce does not work in isolation; it is not a zero-sum game.	Transatlantic S&T cooperation for development has been so far very limited. A concerted approach between donors (e.g. for capacity building) would yield significant gains.	What is lacking is the political will linked with additional financing to forward a sustainable development agenda. Politically charged topics such as global climate change, together with the lack of international adherence to defined goals show need for improvement. EU–U.S. Energy Council. Unfortunately, the climate change deniers seem to be able to cooperate more smoothly than the climate scientists.

(continued)

Table 3.2 (continued)

	Advancing science and technology	Economic effects and competitiveness including innovation	International development *(e.g. in dealing with developementissues in Africa, Asia, etc.)*	Sustainable development *(e.g., environment, energy, peace, etc.)*
What kind of activities are considered useful and needed in order to improve the positive effects of European - U.S. S&T cooperation?	Better mechanisms need to be developed to simplify cooperation (e.g., agency to agency cooperation), and to reduce obstacles including legal and IPR issues or existing fragmentation. Real joint funding mechanisms including incentives for transatlantic cooperation. Better information and dissemination of results. Major cooperation in "big science" projects could be improved, e.g., with respect to global challenges issues, or large-scale research infrastructures where duplication exists and open access could be foreseen/strengthened. Useful approach would be the EU–Framework Program and one of the major U.S. Agencies each allocating budget for joint S&T grand challenges project. Some kind of European–U.S. knowledge database including results of government-funded research projects/programs when there have not been any IP claims.	Matching mechanisms for public and private funds should be encouraged. Commonly agreed standards, open markets for products and services. Better harmonization of export control regulation (ITAR), visa issues. The U.S. and Europe should come together to establish a Bretton Woods for the Innovation Economy that would articulate fair principles of innovation-based competition and combat foreign country mercantilism. EU, the U.S., and Asia should rightly compete, and should compete in high-tech, S&T-based industries, and however, we need better rules to guide this competition. Joint stance from the EU and U.S. toward China on fostering innovation, curbing IPR theft, etc., could be one area of cooperation. There is much to learn from the U.S. about promoting small business/start ups in S&T. Establishment of a transatlantic "market" for research funds/grants to stimulate global excellence. Learning from experiences of university—enterprise cooperation. Foster cooperation to get clean energy technologies into the marketplace, e.g., through infrastructure standards related to electric vehicles and harmonized technical standards as well as testing methods for energy using appliances.	Coordinated policies, approaches, and earmarked budgets on both sides for EU–U.S. cooperation dedicated to international development issues with mandatory participation/support to partners from developing countries. EU and U.S. should co-fund a "grand challenge" pertaining to a developing country issue. There is a desperate need for new approaches in Africa to science, education, application of technology, innovation, development of S&T capacity. Developing a multilateral platform for S&T in support of international development. Grants being made available that target specifically tripartite project/program development (EU, U.S., target country). Joint initiatives at G8/G20 and OECD level. Developer synergies between EC DG Relex/Development and USAID programs. Use of Internet and exchange programs to help build and improve universities in developing countries. Improve planning and coordination among capacity building activities. Creation of a shared enterprise to support S&T for development.	Coordinated policies, approaches, planning, and earmarked budgets on both sides for EU–U.S. cooperation dedicated to sustainable development are necessary. Definition of global goals based on scientific evidence, where EU and the U.S. should act as global leaders. EU and the U.S. should co-fund a "grand challenge" organized around a sustainable challenge. Sharing of best practice models (e.g., energy efficiency) and awareness rising. Clean tech financing. Development of economic models that do not rely on ever increasing consumption as a prime indicator of improved quality of life. Life cycle assessment. Indicators development.

(continued)

Table 3.2 (continued)

	Advancing science and technology	Economic effects and competitiveness including innovation	International development (*e.g. in dealing with developementissues in Africa, Asia, etc.)*	Sustainable development (*e.g., environment, energy, peace,* etc.)
What role is seen for S&T education and mobility?	Maximum mutual openness is needed as this is such an important field to foster competitiveness. Mobility has to include school students, teachers, and researchers at all levels. Europe and the U.S. exchanging experiences in the fields of STEM education and teaching. Maximum exchange of universities in the fields of up-to-date education methods.	Europe can learn from the U.S. when it comes to innovation, entrepreneurship, Venture Capital, start-ups, and its relevance for education and mobility. To instil a business-oriented approach to technology development for university students. Education could—in the long run—help address the risk-averse mentality of the European business and S&T sectors with respective rewards. Increased opportunities for entrepreneurs' mobility. Jobs of Knowledge Economy are needed: currently good training provided for narrow field jobs, more interdisciplinary training is required. Jobs of the future are to be different; however, creating jobs is highly politically influenced that there is not much discourse about S&T/education/mobility creating the knowledge economy of the future.	S&T education should be a corner stone of coordinated strategy. S&T education and mobility are vital for development, including more training opportunities for people/scientists from developing countries and visits of developed countries scientists to developing countries, thus encompassing a new focus in USAID on S&T education and improvement in visa situation. Avoiding creating incentives for brain drain from developing countries.	S&T education should be a corner stone of coordinated strategy. Programs to create awareness about concrete and solvable problems to which joint and practical solutions may be applied based on EU - U.S. planning, know-how, and assets. S&T education and mobility are vital for sustainable development, including more training and exchange opportunities for people/scientists from developing countries as well as improved visa procedures. Creating a large scale program for student exchange covering research in sustainable development.

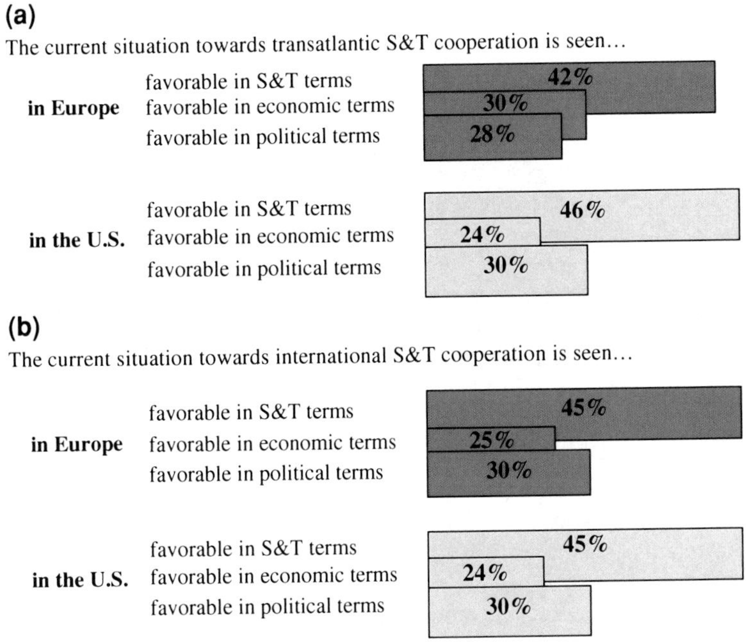

(a)

The current situation towards transatlantic S&T cooperation is seen…

in Europe	favorable in S&T terms	42%
	favorable in economic terms	30%
	favorable in political terms	28%
in the U.S.	favorable in S&T terms	46%
	favorable in economic terms	24%
	favorable in political terms	30%

(b)

The current situation towards international S&T cooperation is seen…

in Europe	favorable in S&T terms	45%
	favorable in economic terms	25%
	favorable in political terms	30%
in the U.S.	favorable in S&T terms	45%
	favorable in economic terms	24%
	favorable in political terms	30%

Fig. 3.1 **a** Current situation toward transatlantic S&T cooperation in Europe and the U.S. specified according to favourable in S&T, economic, and political terms **b** Current situation toward international S&T cooperation in Europe and the U.S. specified according to favourable in S&T, economic, and political terms

transatlantic S&T cooperation in Fig. 3.1a as for international S&T cooperation in Fig. 3.1b.

The distribution of views by U.S. and European respondents was similar, therefore has not been split up in this analysis.

Since this particular question touches upon the very essence of the study, the broader notion of key stakeholders' views and perceptions has been captured by the following spectrum of comments as summarized in Box. 3.1 in response to this question on the situation in Europe and the U.S.

It is interesting to note that implications of the economic/financial crisis are typically seen in terms of budget cuts for S&T, particularly international S&T, as opposed to the option of pooling/coordinating S&T expenditure through international cooperation which has been mentioned only in the fields of space research.

Box. 3.1 Key Stakeholders' Views and Perceptions Captured by Comments in Response to the Question on the Current Situation in Europe and the U.S.

- There is a lot of good political will on all sides, but many "hesitations" when it comes to deliver; everybody tries to impose their own rules and regulations in governance issues.
- As current collaboration is deemed to work pretty well, there is no such high focus on transatlantic cooperation. Both EU and the U.S. are more focused on China, India, Brazil than on each other.
- Europe as a whole is currently more focused on building a European Research Area than outside collaboration. New actions are needed to strengthen transatlantic S&T cooperation.
- Due to the situation of scattered S&T-policies within EU Member States, especially in the definition of global challenges, more "homework" is needed in the EU. European–U.S. cooperation requires a strong EU, which evidently means weakening of the position of individual Member States' bilateral U.S. cooperation to pool resources at EU level.
- The economic and financial crisis has put pressure on governments to limit spending. It also created some tendencies for a more inward-looking approach with respect to S&T support. The situation still remains quite favorable for international S&T cooperation, with a growing focus on addressing global challenges jointly with international partners. Nevertheless, the U.S. remains a priority partner country for S&T cooperation and therefore, transatlantic cooperation was less affected.
- I think Europe is doing a better job than the U.S. of exploring S&T collaborations with Asian countries. Also, I think in general the U.S. has had a tougher time dealing with the competitive threat from Asia than Europe has. Maybe Europe is just more used to international trade, probably its companies have taken a longer term view than the U.S. and not outsourced as much. Probably it's fair to say that EU countries by and large have a better working relationship right now in S&T and trade/economic terms with Asian countries (esp. China) than the U.S. does.
- Tough times seem to bring people together. When goals are too expensive for any one country to obtain, the forcing function to work with others overcomes the reluctance to let other countries into the critical path.
- Within the last years every country has turned more inward looking: on the one hand everyone wants to cooperate more internationally, on the other hand everyone focuses on domestic issues.
- Europe generally seems to have a positive attitude toward both transatlantic and international S&T cooperation. Of course, the current budget realities affect the ability to execute cooperation, but in general Europe seems to remains positive toward the need and opportunities for international cooperation.

- In principle the conditions are favorable; however, people do not think much about Europe but are more attracted to China, India, Brazil, etc.
- The general trend is much stronger toward international S&T cooperation and positively influenced by trade and export considerations. However, there is increasing concern about increasing export of jobs including S&T to India and China.
- The current U.S. administration is highly supportive of S&T cooperation—indeed, this may be the most positive political environment in years. However, the budget realities are such that it is going to be extremely difficult to match the positive attitude with the needed financial resources. Furthermore, the average U.S. taxpayer needs to be convinced that investments in S&T are worth the short-term costs, and that international cooperation is good use of taxpayer dollars.
- Current U.S. politics incorporates an unfortunate nationalistic strain, along with a fiscal conservatism, both of which will militate against funding or encouraging international collaboration in the near term.

Question: What effects have the security-driven considerations in the U.S. had on S&T cooperation with respect to transatlantic European–U.S. S&T cooperation, and international S&T cooperation, respectively?

Figure 3.2 summarizes the views of respondents: 57 % see minor effects motivated by security-driven considerations in the U.S. on transatlantic European–U.S. S&T cooperation, whereas 43 % perceive substantial effects. As for effects on international S&T cooperation, the situation is seen clearly differently: 28 % see minor effects by security-driven considerations in the U.S., whereas 72 % perceive substantial effects.

Asked for other effects resulting from security considerations the most frequently mentioned points are best reflected by the following quote of one respondent:

> Increasing requirements for U.S. visas have negatively affected S&T cooperation throughout the world, and not just S&T but has broader effects. The damage is slow in being recognized but could have a major impact in 10–15 years. The U.S. always received the best and brightest people who wanted to come to the U.S., this might be changing with the changed visa requirements. There is evidence that this has been changing.

Question: Is the EU-Research Framework Program considered an effective approach toward S&T contributions to tackling major challenges and increasing competitiveness? Is it considered an effective approach to fostering transatlantic S&T cooperation?

The European Union's 50 billion Euro "Seventh Framework Programme" is a strategic program fostering cooperative European & International S&T cooperation.

In general, the European Union's Seventh Framework Programme is considered interesting by 78 % of respondents as effective approach toward S&T contribution to major challenges and increasing competitiveness. As demonstrated in Fig. 3.3,

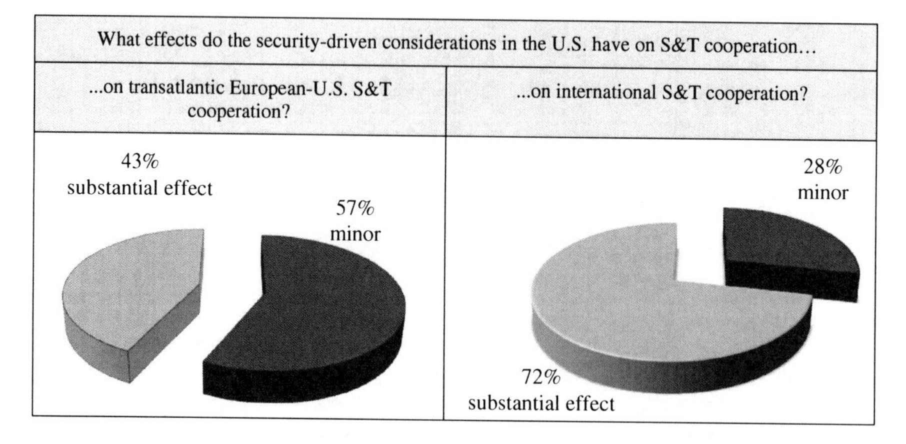

Fig. 3.2 Effects of security-driven considerations in the U.S. on S&T cooperation with respect to transatlantic EU–U.S. S&T cooperation, and international S&T cooperation, respectively

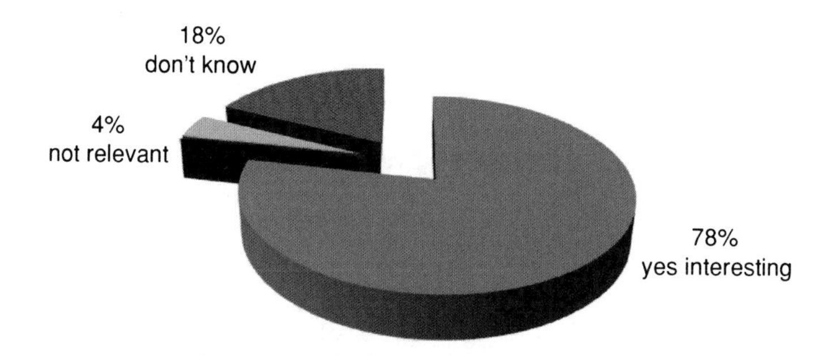

Fig. 3.3 Perceptions of the EU's seventh research framework programme as effective approach toward S&T contributions to tackle major challenges and increasing competitiveness

further 18 % of respondents do not know, 4 % perceive the EU-Framework Program not relevant.

Particularly, the U.S. respondents are regularly impressed by the significant commitment the EU-Framework Program represents as largest, transnational, and cooperative research and technology program providing European tax payers money for research of European and international dimension. While clearly seen as program with strategic mid-term orientation addressing major global challenges and strengthening science-based competitiveness of the EU, the EU-Framework Program has also gained a reputation of being highly bureaucratic with a strong need for simplification.

Asked for the EU Seventh Framework Programme as an effective approach to fostering transatlantic European–U.S. S&T cooperation, 52 % of key stakeholders agree, while 19 % disagree, 23 % do not know, as reflected in Fig. 3.4.

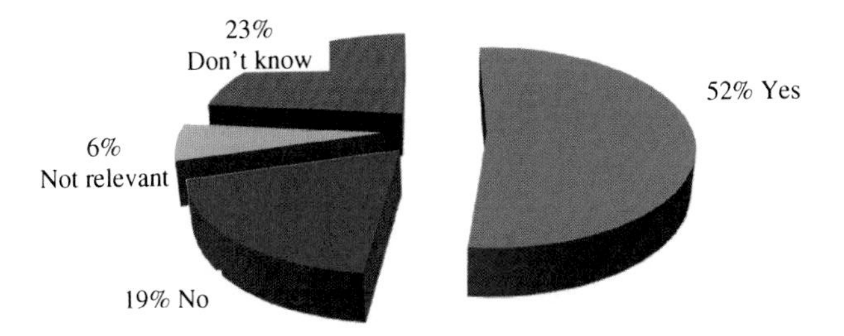

Fig. 3.4 Perceptions of the EU's seventh research framework programme as effective approach to fostering transatlantic EU–U.S. S&T cooperation

More specifically, U.S. participation in the EU-Framework Program seems to lack sufficient awareness and available information on concrete opportunities in the U.S. Several respondents also assume not enough pressure; "need" to get involved due to available resources elsewhere and the bureaucratic efforts involved. As a point of principle it is criticized that the EU-Framework Program applies an "our way or no way" attitude viewed in the U.S. as "arbitrary and contrary to the ethos of a contract-based state activity". This, however, is a point that could be made exactly the other way around, with the difference being that Europeans seem to be used to specific "U.S. requirements".

The bottom-line is that there is way higher potential for transatlantic European–U.S. cooperation as compared to actual U.S. involvement in the EU-Framework Program.

Question: Thematic S&T fields of cooperative activities between Europe and the U.S.: In what thematic fields of S&T are the most promising and substantive opportunities as well as most urgent needs assumed for transatlantic European–U.S. cooperation?

Respondents were asked to indicate the three most important thematic fields in both categories of the question; results are summarized in Table 3.3.

Energy, environment, and climate change are the thematic fields by far perceived with most urgent need for cooperation, while most substantial opportunities for cooperation are seen in health as well as information & communication technologies.

Since replies did not show significantly different trends views of U.S. and European respondents are not split up in this analysis presented here.

Question: Implementing opportunities for transatlantic European–U.S. S&T cooperation: What would be necessary steps?

The majority of key stakeholders seems to be pragmatic and thinks of generating specific thematic initiatives in order to take advantage of opportunities for transatlantic European–U.S. S&T cooperation.

Quite a number of respondents consider the development of shared visions, objectives, and related strategies necessary actions, as well as creating a new policy exchange platform and involving major actors (policy and implementation level). Details plus specifications are summarized in Table 3.4.

Table 3.3 Thematic S&T fields of cooperative activities between Europe and the U.S.: most promising and substantive opportunities as well as most urgent needs for transatlantic European–U.S. cooperation

Thematic field	**Most substantial opportunities for cooperation** Most urgent need for cooperation Percentages indicated refer to key stakeholders replies
Health	**18%** consider Health the most substantial Opportunity **11%** consider Health most urgent
Food, Agriculture, Biotechnology	**10%** consider these fields the most substantial opportunity, in particular Biotech **13%** consider these fields most urgent, in particular Biotech
Information and Communication Technologies (ICT)	**18%** consider ICT the most substantial opportunity **9%** consider ICT most urgent, in particular future internet
Nanotechnologies, Materials, Production technologies	**8%** consider these fields the most substantial opportunity **7%** consider these fields most urgent
Energy	**11%** consider Energy the most substantial opportunity,in particular renewable energy **24%** consider Energy most urgent, in particular energy efficiency and affordable energy for least developed countries
Environment, Climate Change	**11%** consider these fields the most substantial opportunity, in particular Climate Change **24%** consider these fields most urgent
Transport	**7%** consider transport the most substantial opportunity, in particular electrical vehicle and battery development **4%** consider transport most urgent
Social Sciences and Humanities	**6%** consider these fields the most substantial opportunity **4%** consider these fields most urgent, in particular methodologies, surveys, open access to e-libraries
Space, please specify	**5%** consider space the most substantial opportunity **0%** consider space most urgent
Security, please specify	**3%** consider securitythemost substantial opportunity **5%** consider security most urgent

Question: What could be pilot projects and examples for future effective ways of joint European–U.S. S&T efforts?

A number of various thematic pilot projects and examples were suggested covering basically most of the thematic fields with some focus on what is typically considered global challenges issues such as health, sustainable energy, security challenges, etc. In terms of more general suggestions the following have been mentioned frequently:

Table 3.4 Summary of implementing steps for transatlantic European–U.S. S&T cooperation

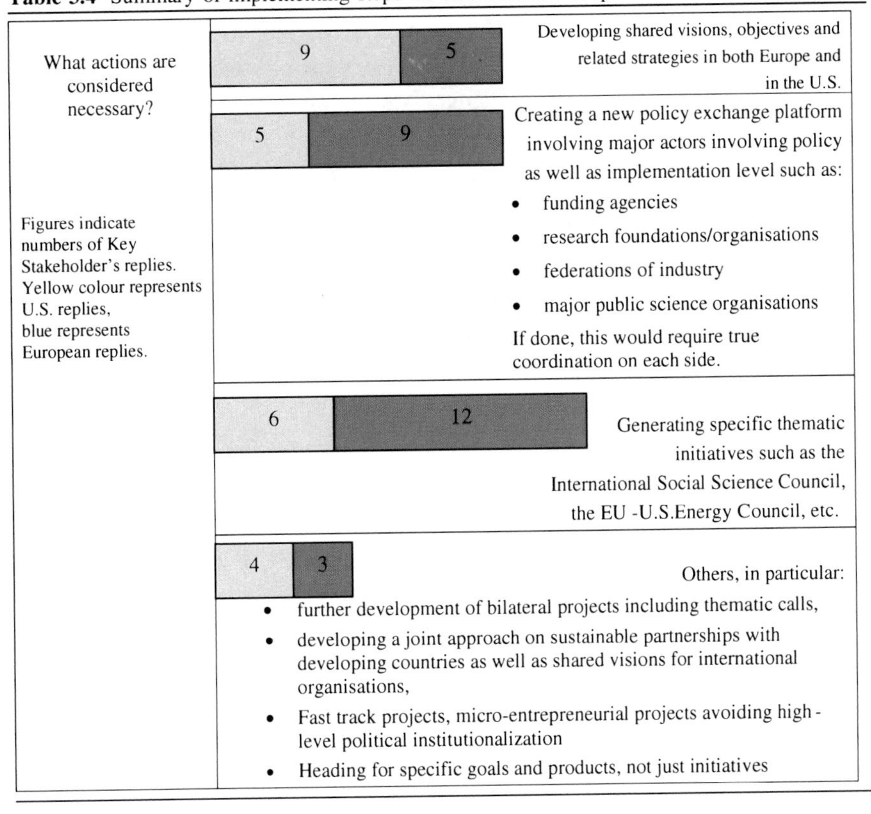

Common procedures in order to overcoming fragmentation and using synergies instead of increasing: common calls for proposals and deadlines, mutual review mechanisms, funding regardless of origin, IPR mechanisms, shared areas in work programs, etc., thus allowing real cooperation without too many hurdles.

Using existing mechanisms that are adapted to transatlantic cooperation needs: specified transatlantic domains could be included in the EU-Framework Programs with streamlined, clearly defined application procedures and deliverables.

Joint initiatives around more general joint interests and values such as protecting intellectual property.

Creating EU–U.S. jobs based on initiatives built on research and development innovations, creating something like transatlantic jobs for the future and transatlantic hi-tech jobs.

Small-scale international science-driven partnerships as models: identification and analysis of suitable models and development of trust-building measures between the various actors used as basis to scale-up effective measures that expand support for these types of activities across agencies and countries.

Specific thematic topics with narrow focus: As the agreement on broader shared visions—thought a laudable goal—typically proof difficult, efforts targeted at highly specific topics might be more successful, examples could be IT data bank for research animal models in biomedical research, etc.

Transatlantic S&T cooperation with international dimension:

- Developing transatlantic cooperative activities in order to tackle international challenges such as global health issues as defined by the NIH-National Institutes of Health.
- S&T capacity building projects in developing countries.
- Large S&T infrastructure projects in energy research areas.
- International forums to foster exchange and coordinate efforts such as IPCC—Intergovernmental Panel on climate change that involves contributions from more than 1,600 scientists.

Cooperation toward the education system: As an effective education system is an issue both in Europe as well as in the U.S. closer cooperation would be useful, thus going beyond the transatlantic "war for talents".

Question: Which actors are perceived as potential driving forces for pushing transatlantic S&T cooperation?

Scientists, scientific organizations, and funding agencies have been considered the most significant driving forces for pushing these activities as detailed in Fig. 3.5.

Question: What specific challenges have been mentioned in taking advantage of opportunities for transatlantic S&T cooperation?

Political interests linked with budgets: Budgets reflect political interests. Thus, money needs to be earmarked for transatlantic S&T cooperation on both sides with a mid- to long-term perspective if there is true political and concrete interest. Clear goals and ways of measuring success are essential. Accordingly, joint projects need to demonstrate their added value and benefits from S&T outcomes including appropriate reward structures for funding bodies.

Focus on the real issues: instead of short-term individual interests. Scientists have to be in the driving seat. In organizational terms, driving forces could be major governmental research organizations in order to ensure excellence, focus on areas of highest priority to S&T policy makers, and access to funding sources.

Rules and procedures: need to be streamlined and harmonized in order to truly foster effective transatlantic S&T cooperation. Following the principle of real partnership governing and funding bodies need to be ready for compromise and step-back from national-driven approaches. A pragmatic balance of accountability and flexibility has to be worked out in order to demonstrate that rules and procedures are intended to form the framework for making cooperation happen.

Information and attitudes: Access to information and opportunities for cooperation has to be made as easy as possible including up-to-date tools for networking and potential collaborators. The analysis of examples for cooperation—both positive and negative—should lead the way: ITER, CERN, international telescopes, etc.

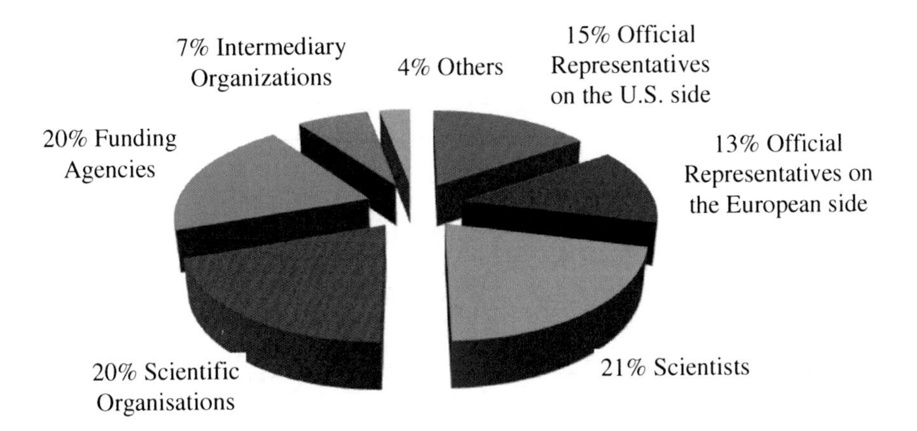

Fig. 3.5 Overview of actors perceived as potential driving forces for pushing transatlantic S&T cooperation

As much as Europe and the U.S. are organized differently, in working on effective ways of transatlantic cooperation actors have to go beyond simplistic prejudices such as "Europe is run more by governments", "the U.S. is more run by companies", "old versus new Europe", etc.

The "*Knowledge Economy*" is a challenge both for Europe and the U.S. likewise. Particularly, in view of emerging economies' developments Europe and the U.S. need to define their future economic and societal models and systems. In working on common goals S&T can contribute cost-effective, cost-shared solutions.

3.2.2 Learning from Experiences

Question: What mechanisms of transatlantic European–U.S. S&T cooperation have worked well and why?

Table 3.5 summarizes key stakeholders' experiences in mechanisms that have worked well for transatlantic S&T cooperation and gives indications on why these mechanisms worked well.

The majority of respondents judge bottom-up non-predefined mechanisms driven by real interest as best working approaches.

Question: What could be examples of mechanisms that worked well and major lessons learned?

Examples with well-working mechanisms mentioned include the EU–U.S. Biotechnology Task Force setup a bit more than 10 years ago, some bilateral intergovernmental programs, IPCC, several mobility program,s and EU–U.S. cooperative activities in Space research. A brief description plus lessons learned is summarized in the following Table 3.6. Of course there are a number of questions to be discussed, such as how to measure "success" of mentioned examples, is

Table 3.5 Mechanisms of transatlantic European–U.S. S&T cooperation that have worked well and reasons why

What mechanisms worked well (%)	Why did mechanisms work well (%)
35 Bottom-up, no predefined mechanisms	43 Real interest
22 Predefined funding mechanisms including joint European–U.S. funding streams	27 Clear objectives, implementation mechanisms
19 Specific thematic mechanisms	14 No predefined conditions
18 Predefined funding mechanism based on individual country-specific arrangements	12 Detailed predefined conditions
6 Other	4 Other

there a higher probability of success of bottom-up versus top-down initiatives. There are no general "rules", lessons learned will represent helpful guidance though. The recently published report on "Knowledge, Networks and Nations: Global scientific collaboration in the 21st century" by the Royal Society also provides interesting analysis and reflections on the issue.[3]

Question: What could be examples of mechanisms that have not worked well and major lessons learned?

Examples with not well working mechanisms mentioned include the EU–U.S. Energy Council, Coordinated Calls between EU and the U.S., the European ScienceFoundation, and the EU-Framework Program. A brief description plus lessons learned is summarized in the following Table 3.7.

Question: What mechanisms of transatlantic European–U.S. S&T cooperation have not worked well and why?

Table 3.8 summarizes key stakeholders experiences in mechanisms that have not worked well for transatlantic S&T cooperation and gives indications on why these mechanisms have not worked well.

The majority of respondents' judge predefined funding mechanisms based on individual country-specific arrangements and a lack of real interest as least well-working approaches.

[3] The Royal Society (2011) Knowledge, Networks and Nations: Global scientific collaboration in the twenty-first century. RS Policy document 03/11.

Table 3.6 Examples of mechanisms that worked well and major lessons learned

Example of well working mechanisms	Description	Lessons learned
EU–U.S. biotechnology task Force	Promotion of information exchange and coordination between biotechnology research programs funded by the European Commission and the U.S. government. The Task Force was originally conceived as a medium that scientists and science administrators could use as a think-tank about the future of biotechnology research. It was seen as means to increase the mutual understanding of the U.S. and European Community activities and programs related to biotechnology research.[a] Setup a bit more than 10 years ago, the setup for suitable mechanisms has been included in the discussions right from the beginning and resulted in the adaptation of existing instruments.	Activities built on mutual political interest and benefits, funding opportunities were considered, close link between science and policy making.
Bilateral intergovernmental programs	Bilateral activities between EU Member States and U.S. Agencies/ Departments, typically focussed on specific topics such as security research, energy research, health, etc.	Activities built on mutual political interest and benefits, budgets made available, close link between science and policy making.
IPCC—intergovernmental panel on climate change	IPCC is the leading international body for the assessment of climate change. It was established by the United Nations Environmental Program and the World Meteorological Organization to provide the world with a clear scientific view on the current state of knowledge in climate change and its potential environmental and socio-economic impacts. Thousands of scientists from all over the world contribute to the work of IPCC on a voluntary basis. International platform for the coordination of climate change objectives, measures with substantial involvement of leading international scientists, agencies, intermediaries, policy makers.[b]	Establishment of international platform with significant impact on global climate change development through coordination efforts and involvement of the relevant parties. Comment on IPCC and its development also available at [c]

(continued)

Table 3.6 (continued)

Example of well working mechanisms	Description	Lessons learned
Mobility/Visiting scholar programs	Programs setup between governments or other organizations in order to foster exchange and mobility of researchers, students, etc. Typically mechanisms of mobility programs are relatively simple and straightforward, not implying heavy bureaucratic efforts.	Eventually cooperation is driven by people, therefore, individuals who participated in mobility programs serve as messengers and bridgeheads to the pragmatic cooperation between institutions. Researchers with international experience are also those most likely to engage in international S&T cooperation.
EU–U.S. cooperative activities in Space research	Cooperative activities in Space research initiated and driven by U.S. government representatives/agencies and EU bodies (ESA—European Space Agency, European Commission). Funding mechanisms are well defined, mechanisms follow the principle of complementarity and keep money flows separate.	Meeting a common science goal while budgetary shortfalls on both sides (in particular on the U.S. side), motivated by the fact that major space research activities are extremely expensive.
EU-Framework Program	Multi-annual European Framework Program for Research, technological Development and Demonstration offering competitive funding for cooperative research projects.	Instrument to providing predefined funding mechanisms including joint European-international funding streams also applicable to transatlantic EU–U.S. cooperation.

[a] Details on the EU - U.S. Biotechnology Task Force: http://ec.europa.eu/research/biotechnology/eu-us-tak-force/index_en.cfm?pg=tf_mission
[b] Details on IPCC: www.ipcc.ch/organization/organization.shtml
[c] The Royal Society (2011) Knowledge, Networks and Nations: Global scientific collaboration in the twenty-first century. RS Policy document 03/11

Table 3.7 Mechanisms of transatlantic European–U.S. S&T cooperation that have not work well and reasons why

What mechanisms did not work well (%)	Why did mechanisms NOT work well (%)
35 Predefined funding mechanism based on individual country-specific arrangements	43 No real interest
22 Predefined funding mechanisms including joint European–U.S. funding streams	27 No clear objectives or implementation mechanisms
19 Specific thematic mechanisms	14 No predefined conditions
18 Bottom-up, no predefined mechanisms	12 No detailed predefined conditions
6 Other	4 Other

Table 3.8 Examples of mechanisms that have not worked well and major lessons learned

Example of mechanisms not working well	Description	Lessons learned
EU–U.S. energy council	Initiated between the U.S. and the European Commission in Nov. 2009, the Energy Council was setup to provide a new framework for deepening the transatlantic dialog on strategic energy issues such as security of supply or policies to move toward low carbon energy sources while strengthening the ongoing scientific collaboration on energy technologies.[a][]	The EU–U.S. Energy Council has not yet had the impact it could have had in terms of real output. Direct involvement of scientists is of clear importance in the agenda setting process. The Energy Council seems to be deeply involved in a vast number of topics and procedures. However, as the U.S. has put in place similar approaches with other major actors including China and India, the U.S.–EU Energy Council is expected to prove its added value and comparative advantages
Coordinated calls between EU and the U.S.	"Coordinated" calls for proposals between Europe and the U.S. try to coordinate topics; they typically apply their respective procedures	Coordinated calls involve parallel reviews and procedures, thus are overly time-consuming and not particularly effective in generating critical results of research projects
EU-Framework Program	Multi-annual European Framework Program for research, technological development, and demonstration offering competitive funding for cooperative research projects. As much as the EU-Framework Program is open to international partners, the participation of U.S. organizations—though Nr. 1—is far from its potential	The ability of the U.S. organizations to participate in the EU-Framework Program is somewhat limited due to certain legal constraints and mismatching approaches of the U.S. Agencies

[a] Details on EU–U.S. Energy Council: http://europa.eu/rapid/pressReleasesAction.do?reference=IP/09/1674

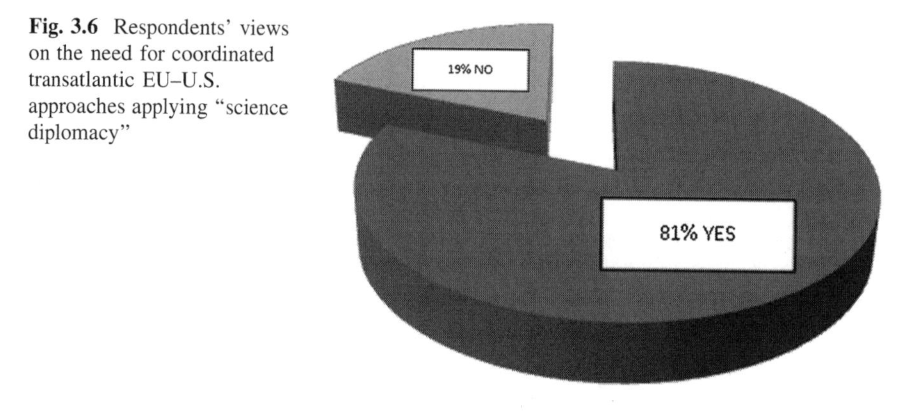

Fig. 3.6 Respondents' views on the need for coordinated transatlantic EU–U.S. approaches applying "science diplomacy"

3.2.3 Science Diplomacy and European–U.S. Cooperation

The AAAS-American Association for the Advancement of Science[4] as international non-profit organization dedicated to advancing science around the world, it serves more than 260 affiliated societies and academies of science and 10 million individuals. AAAS fulfills its mission through initiatives in science policy, international programs, science education, and more.

AAAS defines science diplomacy as "international science cooperation to foster communication and cooperation among the peoples of diverse nations and to promote greater global peace, prosperity and stability".

The term "Science Diplomacy" is differently used with respect to:

- "Science for diplomacy": to facilitate relations with other countries with a view of expanding collaborations and exchanges in areas such as trade, to attract investments, or to promote stability.
- "Diplomacy for science": reaching out to other countries for knowledge sourcing, for pooling resources to address jointly big S&T-based challenges, share costs, etc.

Rhetoric and real action with respect to science diplomacy are very different in Europe and in the U.S.

Question: Is there a need for coordinated transatlantic European–U.S. approaches applying "science diplomacy"?

About 81 % of respondents as displayed in Fig. 3.6 think there is a need for coordinated transatlantic European–U.S. science diplomacy approaches, with a slightly higher share of the U.S. as compared to European respondents.

It is interesting to note that Europe does apply a broad range of science diplomacy approaches for instance via its EU-Framework Program, however, does not perceive these activities as such.

---

[4] Information on AAAS at www.aaas.org.

Examples for transatlantic science diplomacy activities as mentioned could be:

- Topics with political and economic implications such as global health, emissions and climate change, security, etc.
- Concerted efforts toward Third Countries, particularly in countries where the EU and the U.S. already put forward activities (for instance DPRK-Democratic People's Republic of Korea, Iran)
- Developing joint positions for international forums (for instance G8/20 initiatives)
- Joint activities for Africa
- Developing joint approaches in dealing with crisis, for instance via setting up EU–U.S. decadal planning forums in a number of scientific disciplines

Question: What could be examples for existing coordinated transatlantic approaches applying "science diplomacy" or approaches that touch upon "science diplomacy"?

About 50 % of key stakeholders are aware of examples for existing coordinated transatlantic science diplomacy approaches. Some of the examples are listed as follows:

- NIH-National Institutes of Health together with the European Commission joint research efforts on rare diseases.
- The International Science and Technology Centers (ISTC) in Moscow: The center was setup after the collapse of the former Soviet Union to address the risk of proliferation of expertise particularly in the fields of weapons of mass destruction. A centre with the same objective was setup in Ukraine, however, with different membership. Although science diplomacy was not part of the original objectives of the centers, it is clear that the activities of the centers contributed to develop and facilitate relationships with the then newly established Republics. Organized as multilateral organizations, the core members (EU and U.S.) worked together and coordinated some of their actions through broad discussions.
- The International Space Station: it has been justified almost entirely on its contribution to improving relations with the former USSR nations, especially with Russia. Cooperation between countries on space-based science investigations brings together divers peoples together.
- Working together on poverty-related diseases in Africa.
- Activities of AAAS and the Royal Society.
- The Fulbright Program has been setup exactly with the scope of strengthening among others, the S&T dialog of various peoples with the U.S.

Question: Are there examples for "science diplomacy" approaches NOT related to Europe and the U.S.?

About 65 % of respondents are aware of science diplomacy approaches not related to Europe and the U.S. as demonstrated in Fig. 3.7.

Examples given include:

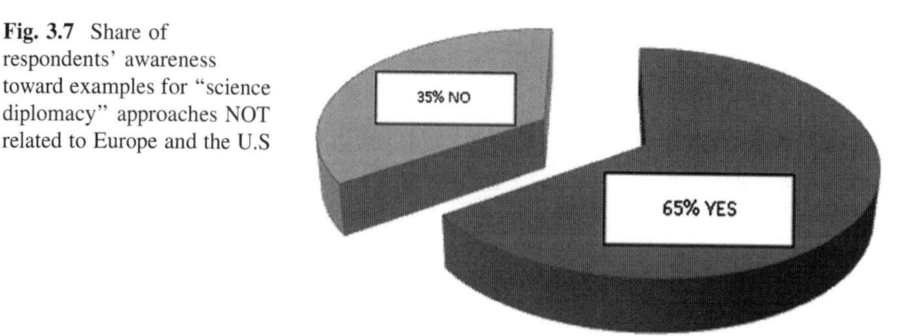

Fig. 3.7 Share of respondents' awareness toward examples for "science diplomacy" approaches NOT related to Europe and the U.S

- Education- and research-related issues are always among the first activities to build/strengthen contacts
- Visiting Scholar Programs
- U.S.–China Clean Energy Research Center (CERC)
- World Bank efforts toward building S&T capacities in the least developed countries
- The U.S. State Department program to enhance science diplomacy with the Middle East
- NIH's Fogarty Institute toward Africa
- U.S.–India energy partnership and U.S. Asia–pacific partnership on energy and climate change
- CRDF Global activities are largely oriented toward science diplomacy, including U.S.–DPRK Science Engagement Consortium, work with the National Academy of Sciences in Syria, ongoing development of possible initiatives with Africa.

Question: What would be necessary in order to develop coordinated transatlantic science diplomacy activities?

Diplomacy normally serves the interests of a state, e.g., the U.S. tries to build on its positive image as a science nation to increase its standing in the Muslim world.

For coordinated transatlantic science diplomacy activities to make sense and be effective the majority of respondents suggested that it is needed to generate a clear understanding of the shared objectives between Europe and the U.S.

This could be achieved by setting up a mandated group of experts from the U.S. and Europe, tasked to develop shared objectives, a strategy, priority regions, and a set of recommendations for implementing steps with respect to what activities, where, resources, actors, timing,and desirable outcomes. The result could be the basis for a well-defined cooperative program between Europe and the U.S.

3.3 European–U.S. Science and Technology Cooperation from a Systems Sciences Point of View

According to Wikipedia, systems science is

> an interdisciplinary field of science that studies the nature of complex systems in nature, society, and science. It aims to develop interdisciplinary foundations, which are applicable in a variety of areas, such as engineering, biology, medicine and social sciences. Systems sciences covers formal sciences fields like complex systems, cybernetics, dynamical systems theory, and systems theory, and applications in the field of the natural and social sciences and engineering, such as control theory, operations research, social systems theory, systems biology, systems dynamics, systems ecology, systems engineering and systems psychology.

In the context of this study, systems sciences is considered highly relevant as transatlantic S&T cooperation, particularly when it comes to issues of strategic importance, takes place in a complex system. Decisions taken do not happen in isolation but are interconnected with the developments in the wider environment of Science, Technology, Innovation, Economy as well as Policy.

These factors create a system with its own behavior and rules. It is shaped by the various actors, networks, and dynamics in the system.

From the systemic point of view, specific behavior and actions create reactions with broader impact on all players and the functionality in the entire system.

So far, this systemic view is—at best—hardly taken into account, as demonstrated by the various actors with all their specific activities. Transatlantic European–U.S. S&T cooperation has to be more than just the well-established routine interaction of official representatives and "usual suspects" actors. Instead it has to be understood as an interconnected system of actors which is influenced by each actor's activities, thus touching upon systems sciences and cybernetics.

This study could not go into the systemic details; however, the author is convinced that a more systemic view and related approaches will be needed for future increased effectiveness of actions at all levels. As a starting point, it would be worthwhile to explore adequate systems sciences methods in order to analyze the current status of transatlantic S&T cooperation from a systems sciences point of view.

Chapter 4
Overall Conclusions and Policy Recommendations

Abstract The purpose of this study was to identify major trends in key stakeholders' views and perceptions on strategic transatlantic S&T cooperation. As a result, the following overall issues turned out to be of overarching importance and relevance. They represent the essence derived from the various sources of input generated for this study, and are summarized as seven conclusions and three blocks of policy recommendations. Each of the policy recommendations has been split into a recommendation for strategic development and the respective recommendation for action in order to provide guidance for implementation.

4.1 Overall Conclusions

The purpose of this study was to identify major trends in key stakeholders' views and perceptions on strategic transatlantic S&T cooperation.

As a result, the following overall issues turned out to be of overarching importance and relevance. They represent the essence derived from the various sources of input generated for this study, and are summarized as central conclusions complemented with the authors' interpretation.

Put together in one phrase, the overall conclusion could be summarized by saying that transatlantic S&T cooperation has become a job, it is not that much driven by a mission and/or vision contributing, to solve major challenges.

S. E. Herlitschka, *Transatlantic Science and Technology*, SpringerBriefs in Business,
DOI: 10.1007/978-1-4614-4385-8_4,

4.1.1 Conclusion 1: The Economic Reality: Europe and the U.S—Biggest Economic Players, Not Perceived and Appreciated as Such

Europe and the U.S. are the biggest players in economic terms—still—and have the highest level of expertise to contribute in dealing with global challenges. Their economies are heavily interconnected and dependent on each other. However, this fact is not commonly recognized and appreciated as such.

The current financial and economic crisis on both sides of the Atlantic demonstrates the interconnectedness as well as the need for coordinated actions. Instead, it seems to trigger the trend toward more inward orientation and inward looking—both in Europe and in the U.S.—which impacts various fields, together with the notion to compete for the smaller economic problem.

Particularly when it comes to the transformation to what is typically referred to as "knowledge economy", Europe and the U.S. could clearly benefit from each others' experiences and efforts, since the "knowledge economy" poses huge challenges particularly to highly developed economies with its implications in terms of future job profiles, the practical relevance of "open innovation" and the innovation system, the future role of established capabilities such as manufacturing, and the role of science, technology, and innovation.

4.1.2 Conclusion 2: Complexity of Systems Perceived Likewise, Adds Complexity at Instruments Level and Creates Unattractive Framework for Cooperation

Each respective system, Europe as well as the U.S. (and other systems), is perceived by the other side as highly complex and hard to understand. Repeatedly, key issues such as the variety of actors, decentralization versus coordination, specific politically motivated interest groups, short-term interests, etc. have been mentioned.

This complexity of each system is also reflected by the instruments set up toward transatlantic S&T cooperation, thus adding up to multiple layers of complexity. For participants or potential participants in transatlantic S&T cooperation this situation generates highly unattractive framework conditions.

4.1.3 Conclusion 3: Diverging Driving Interests and Priorities Rather Separate than Unify Europe and the U.S., Despite Pressing Global Challenges

Despite similar societal values, Europe and the U.S. are driven by fundamentally diverging forces: competition versus cooperation, short-term versus long-term orientation, knowledge generation versus business generation, differing views on the relevance of security, etc.

Nevertheless, Europe and the U.S. are confronted with joint global challenges too big to deal with independently. And both systems are massively interconnected also in succeeding or failing when it comes to dealing with our joint global challenges.

4.1.4 Conclusion 4: Shared Values Have Not Yet Led to Comprehensive Joint Visions and Coordinated Strategies

There are a number of issues on which Europe and the U.S. work together, based on common interests and an expedient alliance type of relationship. Europe and the U.S. are those regions closest in terms of shared values.

In the light of major global challenges, a joint vision, real joint ambition, and corresponding strategies based on shared values and appreciation—though important—are in question or not existing.

4.1.5 Conclusion 5: "Typical": Stereotypes, Misperceptions, and Prejudices Still Exist

Typical, simplistic stereotypes exist and are found in the public as well as in wide parts of the S&T community.

Examples such as "Europe is just complicated", "with Europeans you always get lost in procedures", "Europeans are busy with themselves and their internal procedures", as well as "Americans are simplistic", "Americans are interested in short-term opportunities only", "America is driven by lobbyists' interests instead of concentrating on real issues", etc. are surprisingly widely spread.

4.1.6 Conclusion 6: International Orientation and the "Hype" for China: Very Different Approaches

The perception of the purpose and benefits of real international S&T cooperation is fundamentally different on both sides. Also, in the case of international cooperation, it is true that money follows strategy, assuming that where there is no money there is no interest.

The U.S. currently seems to be driven primarily by the "hype" for China, which has to be seen in the context of both, the economic dimension and the huge U.S. financial dependence on China, as well as the sometimes naively perceived short-term business opportunities in China.

Europe, so far less interconnected and dependent on China, seems to implement more long-term strategies in international S&T cooperation. Particularly, the EU-Research Framework Program represents a long-term financial commitment to S&T cooperation, including and providing funding for international cooperation outside Europe such as partners in China.

4.1.7 Conclusion 7: Experiences in Strategic Transatlantic Cooperation: Existing Mechanisms are Unsatisfying, New Approaches are Needed

Many key stakeholders in Europe and the U.S. are either unaware of the need for strategic transatlantic S&T cooperation or unsatisfied with the way strategic transatlantic S&T cooperation is currently organized.

The traditionally applied procedures for transatlantic cooperation particularly in areas of strategic importance no longer seem to fit in terms of approach, timing, content, and dynamic. Many of the same hurdles still exist as many years before. Effective ways of strategic cooperation need effective mechanisms to make the cooperation work and deliver results. This clearly has to go beyond the majority of established routines.

Some new initiatives are on the way: efforts in the context of transatlantic economic and innovation-related discussion seem to be gaining momentum, such as the Transatlantic Business Dialog (TABD) and the Transatlantic Economic Council (TEC). Other promising examples are more substantial initiatives in the area of energy as well as "StarMetrics".

Table 4.1 Summary of policy recommendations as derived from the conclusions of the study

Policy recommendations	Overall conclusions
Where to go together: Real Leadership needed	Conclusion 1: The economic reality: Europe and the U.S.–biggest economic players, not perceived and appreciated as such. Conclusion 2: complexity of systems perceived likewise, adds complexity at instruments level and creates unattractive framework for cooperation. Conclusion 3: diverging driving interests and priorities rather separate Europe and the U.S. despite pressing global challenges. Conclusion 4: shared values have not yet led to comprehensive joint visions and coordinated strategies. Conclusion 5: "typical": stereotypes, misperceptions and prejudices still exist.
Bold actions needed: think big and pragmatic	Conclusion 7: experiences in strategic transatlantic cooperation: unsatisfying mechanisms and new approaches on the way.
Think global: Europe and the U.S. toward international cooperation	Conclusion 6: international orientation and the "hype" for China: very different approaches.

4.2 Three Policy Recommendations

Based on the findings of this study—the multidimensional "picture" generated—and overall conclusions, the following three policy recommendations have been developed:

- Where to go together: real leadership is needed
- Bold actions needed: think big and pragmatic
- Think global: Europe and the U.S. toward international cooperation

Each of the policy recommendations has been split into a recommendation for strategic development and the respective recommendation for action in order to provide guidance for implementation.

Table 4.1 summarizes the policy recommendations as they are derived from the overall conclusions of the study, followed by a more in depth description of the recommendations.

4.2.1 Recommendation 1: Where to Go Together: Real Leadership Needed!

Recommendation for strategic development:

"Real Leadership is needed", particularly in political terms, an appeal used in an inflationary way. However, this is what is necessary indeed. The relationship between Europe and the U.S. is built on shared values and similar societal models. From the perspective of Europe or the U.S. likewise, no other region/country is closer with respect to fundamental societal values. Thus, the nature of the transatlantic cooperation is and should be more than just an expedient alliance, but an alliance of joint societal values. It requires active dedication, the agreement on a clear and committing vision for shaping the future developments that go beyond short-term advantages. Joint ambitious objectives should be defined including deviated strategies plus explicit tangible results to be achieved.

In essence, what is needed is the further transition of this transatlantic alliance originally set up against a joint enemy, toward an alliance for the development of joint solutions. The Atlantic Basin Initiative[1] can serve as a good example.

Recommendation for action:

The economic and financial crisis could be instrumental in concentrating on developing a joint vision implemented by coherent strategies. This is a clear opportunity driven by importance and urgency. Activities under the TABD and the TEC have gained substantial momentum. Specifically, strategic S&T should play a major role in order to foster innovation. S&T as part of the wider innovation/competition-related efforts are a corner stone of what is typically referred to as "knowledge economy" and is a developmental challenge both for Europe as well as the U.S. Close links between TABD, TEC, related activities, and S&T should be actively used and expanded.

Transatlantic S&T cooperation does not take place in isolation but needs to be an integrated part of a wider system. It is shaped by the various factors, networks, structures, and dynamics in the system. From the systemic point of view, specific behavior and actions create reactions with broader impact on the players in the system. It is recommended to strengthen the understanding of transatlantic S&T as a common "system" with its interdependencies, taking advantage of methods available ("reality space", simulation/scenarios, organizational constellations, etc.) in order to demonstrate development options and proactively shape them. What is not useful is the combination of the complexity on both sides, Europe and the U.S. thus adding complexity. Instead, the systemic perspective facilitates the setup of a more effective approach for a common transatlantic S&T system.

The currently established structures/mechanisms for transatlantic European–U.S. S&T cooperation reflect the complexity of European and the U.S. systems. These structures and mechanisms have been the same in nature for many years, sometimes even decades, with the same known hurdles in place. It is recommended to deal with mismatching structures and development of adequate solutions geared at generating results effectively.

How can Europe and the U.S. find common grounds based on diverging driving forces? Can Europe and the U.S. take advantage of these differences? Diverging

---

[1] European commission (2011) The atlantic geopolitical space: common opportunities and challenges.

driving forces in Europe and the U.S. can be perceived as hindering factors, but likewise also as opportunities for developing complementary advantages. It is recommended to go beyond stereotypes and simplistic prejudices, each side could contribute its key features, thus being able to foster complementary advantages, such as pragmatic, highly effective structures and procedures for strategic transatlantic S&T cooperation.

Competition has many advantages, overall it has positive impacts on the distribution of resources. As far as Europe and the U.S. are concerned, there is clearly a tension of competition and cooperation, summarized best by the term "Coopetition" (Cooperation and Competition), demonstrating that Europe and the U.S. are competitors in some areas, and cooperation partners in others. It reflects a delicate balance to be taken care of and specifically considered in effectively working on solutions toward global challenges.

One of the most important examples is the competition for the best "brains". While many countries have recognized that the "human factor" will be the essential "competitive factor" for successful future development paths, they have setup specific measures to attract the best people. This is clearly one of the key issues of strategic S&T importance. It is recommended that Europe and the U.S. develop coordinated approaches for "brain circulation", free movement of scientists and engineers in this transatlantic European–U.S. S&T system.

The EU Member States together with the European Commission have setup the "Strategic Forum for International Cooperation" (SFIC) as a formation with the objective to "facilitate the further development, implementation and monitoring of the international dimension of ERA. In practice, this means sharing information and consultation between the partners (Member States and the Commission) with a view to identifying common priorities which could lead to coordinated or joint initiatives. The group also aims at coordinating activities and positions vis-à-vis third countries and within international fora".[2]

SFIC could—in principle—play an important role in coordinating Member States and European Commission activities with respect to transatlantic strategic S&T cooperation. However, there needs to be a clear sense for urgency, real action, and more short-term timing with respect to envisaged activities.

4.2.2 *Recommendation 2: Bold Actions Needed: Think Big and Pragmatic!*

Recommendation for strategic development:

Time has come for bold and pragmatic actions between Europe and the U.S. if global challenges are to be dealt with effectively. Therefore, focus is needed on major issues of concern and pressing global challenges.

2 http://www.era.gv.at/space/11442/directory/11622.html.

Recommendation for action:

Sustainable energy supply and energy use are among the top priorities, for which bold plans and implementing measures should be developed including all relevant actors. Timely and pragmatic ways toward effective solutions and achieving real results should be developed.

Fields/themes of activities should be aligned across all relevant policy areas (in the EU for instance including all responsible directorates general, such as enterprise, trade, internal/security, external, energy, etc.) and "routine" procedure of cooperation toward tangible results and delivery of effective mechanisms should be developed.

The direct involvement of scientists and engineers is an essential element in providing the passion and pressure for results. This factor should be reinforced.

Systematic analysis of significant examples of joint transatlantic S&T cooperation as well as the identification of successful examples is highly recommended in order to foster structured learning. The EU–U.S. Energy Council would be a highly relevant learning case, particularly with respect to examples in the energy field targeted at other regions. Some new initiatives are on the way and should be checked explicitly for the effectiveness of their mechanisms for implementation.

Information is key, information on opportunities of transatlantic European–U.S. S&T cooperation should be made available more broadly, preferably with the help of established, highly renowned, thematically oriented networks in Europe and in the U.S.

4.2.3 Recommendation 3: Think Global: Europe and the U.S. Toward International Cooperation

Recommendation for strategic development:

In effectively dealing with global challenges, Europe and the U.S. should include the international dimension, thus working toward developing a truly global S&T cooperation and funding instrument.

Recommendation for action

This new global program would have to build on contributions from all countries participating in terms of content, mechanisms,, and financing. The EU-Framework Program for Research, Technological Development and Demonstration can serve as example, as it represents the largest transnational cooperative and competitive research program worldwide, open to the entire world.

These efforts toward international cooperation should be complemented with coordinated activities across all policy areas, in particular external relations.

Science diplomacy is an effective way of fostering international cooperation. Opportunities for coordinated science diplomacy measures resulting from transatlantic initiatives should be systematically identified and analyzed for implementation.

Annex

Annex 1: Stage 1 Questionnaire

Transatlantic European–U.S. Science & Technology Alliance working jointly on Grand Challenges: Vision or chance for realization?

Hypothesis

- The substantial contribution of the EU and the U.S. is required in order to deliver results on the global challenges
- Successful transatlantic European–U.S. cooperation in S&T is advantageous for the actors involved and essentially required in dealing with global challenges.

Questions Based on Mentioned Hypothesis

- Do you agree with the hypothesis mentioned? Or do you rather think that there is a different attitude which makes the EU believe stronger in cooperation whereas the U.S. rather believes in competition as major success factor? Should other approaches/target partners be chosen, e.g EU and Asia, etc.?

- What are the most important questions and expectations from your point of view regarding EU–U.S. S&T cooperation?

S. E. Herlitschka, *Transatlantic Science and Technology*, SpringerBriefs in Business,
DOI: 10.1007/978-1-4614-4385-8, © Sabine E. Herlitschka 2013

- Which factors so far have promoted transatlantic EU–U.S. S&T cooperation, which have been limiting factors or hurdles?

- How do you comment the observation that the EU with the European research area has been gradually opening up for structured international S&T cooperation (incl. provision of funding for international partners), whereas in the U.S. truly coordinated international S&T cooperation has not yet come to play a substantial role, which can also be seen by continued U.S. reluctance to provide dedicated funding for international S&T cooperation. This development has been reinforced by the ever increasing security concerns plus related regulations.

Annex 2: Stage 2 Questionnaire

Transatlantic European–U.S. Science & Technology

Opportunities for real cooperation under the present conditions?
Fulbright Schuman Scholarship, Dr. Sabine Herlitschka
The following introduction provides background information on the Project, its focus and objectives, the expected results, and the methodology applied.

What is It About—Summary of the Project

In dealing with major global challenges such as climate change or energy issues, Science & Technology (S&T) is an essential contributor and international Co-operation in S&T has become an imperative. Despite substantial political will in both Europe and the U.S. to respond to this imperative, there are also major hindrances to their doing so.

This project will explore new opportunities and mechanisms for increased transatlantic European–U.S. S&T cooperation under the present conditions of financial- as well as security-driven limitations in the U.S. even as there is increasing openness of the European Research Area. Examples will be presented of effective joint efforts. The focus of this project is not on initiatives of individual researchers, but on more structured and systematic approaches to European–U.S. S&T cooperation.

The Elliot School of International Affairs at George Washington University is hosting this project.

Is It Just About Europe and the U.S.?

While the project is focused on European–U.S. relationships, it can also address issues of international Science & Technology cooperation where both Europe and the U.S. are responding or should/could respond to joint opportunities in other parts of the world, e.g. Africa, Asia, etc.

Who is "Europe"?

In this project "Europe" refers primarily to activities of the European Union (EU) and its activities in Science & Technology at the European Union level. The position of individual European Union Member States is taken into consideration to some extent, although not in a comprehensive way.

What will Be Done: A Qualitative Study and Policy Recommendations

This Fulbright Project will be implemented as a qualitative study based on opinions, views, and expectations of leading stakeholders and practitioners in the field of transatlantic European–U.S. Science & Technology Cooperation. As a result of the study, policy recommendations will be presented to the European Commission, contributors and other interested parties.

Methodology Applied

The project will be carried out in three stages. As a first step, some basic hypotheses were tested, and relevant questions on European–U.S. S&T cooperation were identified. A key element of the second step will be to conduct a Survey with about 50 leading European and U.S. stakeholders/experts in the field. The third step will be to assemble a smaller group of experts who review the results of steps one and two.

1. Role and opportunities of European–U.S. S&T cooperation
Science and Technology (S&T) plays an ever increasing role in today's changing world of emerging economies, global challenges, global networking, etc. Europe and the U.S. jointly represent a partnership of app. 60 % of world GDP which provides great opportunities but also creates responsibilities for action.

What role and effects do you see for European–U.S. Science and Technology (S&T) cooperation in the following four specific categories: advancing S&T, economic effects and competitiveness, international development, and sustainable development?

Please select options by inserting an "x" into the brackets.

() = single choice

	Advancing Science & Technology	Economic effects and competitiveness including innovation	International development (e.g. in dealing with development issues in Africa, Asia, etc.)	Sustainable development (e.g. Environment, energy, peace, etc.)
How strong do you see the current effects of European–U.S. S&T cooperation in the following areas listed in this line.	() no effects () slight effects () substantial effects	() no effects () slight effects () substantial effects	() no effects () slight effects () substantial effects	() no effects () slight effects () substantial effects
In which areas listed in this line do you see a need for improvement of S&T effects and thus respective activities?	() no improvements needed () slight improvements needed () substantial improvements needed	() no improvements needed () slight improvements needed () substantial improvements needed	() no improvements needed () slight improvements needed () substantial improvements needed	() no improvements needed () slight improvements needed () substantial improvements needed
Could you give examples?	Please specify	Please specify	Please specify	Please specify
What kind of activities would you consider useful and needed in order to improve the positive effects of European–U.S. S&T cooperation?	Please specify	Please specify	Please specify	Please specify
What role do you see for S&T education and mobility?	Please specify	Please specify	Please specify	Please specify
Further comments?	Please specify	Please specify	Please specify	Please specify

How would you describe the current situation in Europe and the U.S. in S&T, political and economic terms towards transatlantic as well as international S&T cooperation?

Please select options by inserting an "x" into the brackets.

[ ] = multiple choice

	Current situation towards transatlantic S&T cooperation is...	Current situation towards international S&T cooperation is...	Please specify your view, if possible give examples
Europe	[] favorable in S&T terms [] favorable in economic terms [] favorable in political terms [] not favourable in S&T terms [] not favorable in economic terms [] not favorable in political terms [] no influence	[] favorable in S&T terms [] favorable in economic terms [] favorable in political terms [] not favourable in S&T terms [] not favorable in economic terms [] not favorable in political terms [] no influence	Please specify
U.S.	[] favorable in S&T terms [] favorable in economic terms [] favorable in political terms [] not favourable in S&T terms [] not favorable in economic terms [] not favorable in political terms [] no influence	[] favorable in S&T terms [] favorable in economic terms [] favorable in political terms [] not favourable in S&T terms [] not favorable in economic terms [] not favorable in political terms [] no influence	Please specify

In your view, what effects have the security-driven considerations in the U.S. had on S&T cooperation...

Please select options by inserting an "x" into the brackets. () = single choice

On transatlantic European–U.S. S&T cooperation?	On international S&T cooperation?	If you think of other effects please specify
() no effect () minor effect () substantial effect Please specify	() no effect () minor effect () substantial effect Please specify	Please specify Please specify

The European Union with its Flagship Program—the 50 billion Euro "7th Framework Programme"—is a strategic program fostering cooperative European & International S&T cooperation.

In general, do you consider the European Union 7th Framework Programme an effective approach towards S&T contributions to tackling major challenges and increasing competitiveness?	() Yes, interesting, because. () No, not interesting because. () Not relevant, because. () Don't know	Please specify

(continued)

(continued)

Do you consider the European Union 7th Framework Programme an effective approach to fostering transatlantic European–U.S. S&T cooperation	() Yes () No () Not relevant () Don't know	Please specify

Thematic S&T fields of cooperative activities between Europe and the U.S.

In what thematic fields of S&T do you see the most promising and substantive opportunities as well as most urgent needs for transatlantic European–U.S. cooperation?

Please check the three most important fields according to the following table.

Please select options by inserting an "x" into the brackets.

[ ] = multiple choice, max. three

Thematic field	Most substantial opportunities for cooperation	Most urgent need for cooperation
Health, please specify	[ ]	[ ]
Food, Agriculture, Biotechnology, please specify	[ ]	[ ]
Information and Communication Technologies, please specify	[ ]	[ ]
Nanotechnologies, Materials, Production Technologies, please specify	[ ]	[ ]
Energy, please specify	[ ]	[ ]
Environment, Climate Change, please specify	[ ]	[ ]
Transport, please specify	[ ]	[ ]
Social Sciences and Humanities, please specify	[ ]	[ ]
Space, please specify	[ ]	[ ]
Security, please specify	[ ]	[ ]
Other, please specify	[ ]	[ ]

In taking advantage of opportunities for transatlantic European–U.S. S&T cooperation…

Please select options by inserting an "x" into the brackets. [ ] multiple choice

...what actions do you consider necessary?	[ ] Developing shared visions, objectives and related strategies in both Europe and in the U.S. [ ] Creating a new policy exchange platform involving major actors involving policy as well as implementation level such as, please specify... [ ] Generating specific thematic initiatives such as the International Social Science Council, the EU–U.S. Energy Council, etc. [ ] Others, please specify...
...what could be pilot projects/examples for future effective ways of joint European–U.S. S&T efforts?	Please specify
...who do you think should be the driving force for pushing these actions?	[ ] Official representatives on the U.S. side, such as.... [ ] Official representatives on the European side, such as.... [ ] Scientists [ ] Scientific Organisations [ ] Funding Agencies [ ] Intermediary Organisations [ ] Others, please specify...
...where do you see specific challenges?	Please specify

2. Learning from experiences

Based on your knowledge or experience, what mechanisms of transatlantic European–U.S. S&T cooperation have worked well...

Please select options by inserting an "x" into the brackets. [ ] multiple choice

What mechanisms worked well?	Why did mechanisms work well?	Examples of mechanisms that worked well?	What do you consider to be the major lessons learned from mechanisms that worked well?
[ ] bottom-up, no predefined mechanisms [ ] specific thematic mechanisms [ ] predefined funding mechanisms including joint European–U.S. funding streams [ ] predefined funding mechanisms based on individual country specific arrangements [ ] other, please specify	[ ] no predefined conditions [ ] detailed predefined conditions [ ] real interest [ ] clear objectives, implementation mechanisms, etc. [ ] other, please specify	Please specify	Please specify
Please specify	Please specify		

Based on your knowledge or experience, what mechanisms of transatlantic EU–U.S. S&T cooperation did NOT work well…

What mechanisms did not work well?	Why did mechanisms not work well?	Examples of mechanisms that did not work well?	What do you consider to be the major lessons learned from mechanisms that did not work well?
[ ] bottom–up, no predefined mechanisms [ ] specific thematic mechanisms [ ] predefined funding mechanisms including joint European–U.S. funding streams [ ] predefined funding mechanisms based on individual country specific arrangements [ ] other, please specify…. Please specify	[ ] no predefined conditions [ ] detailed predefined conditions [ ] no real interest [ ] no clear objectives, implementation mechanisms, etc. [ ] other, please specify Please specify	Please specify	Please specify

3. Science diplomacy and European–U.S. cooperation

The AAAS defines science diplomacy as "international science cooperation to foster communication and cooperation among the peoples of diverse nations and to promote greater global peace, prosperity and stability". However, rhetoric and real action with respect to science diplomacy are very different in Europe and in the U.S.

Please select options by inserting an "x" into the brackets. () = single choice

Do you see the need for coordinated transatlantic European–U.S. approaches applying "science diplomacy"?	( ) Yes ( ) No	Please give examples where you see a need for coordinated European–U.S. approaches of science diplomacy
Are you aware of examples for existing coordinated transatlantic approaches applying "science diplomacy" or approach that touch upon "science diplomacy"?	( ) Yes ( ) No	Could you give examples of coordinated transatlantic approaches applying "science diplomacy" or approach that touch upon "science diplomacy"?
Are you aware of examples for "science diplomacy" approaches not related to Europe or the U.S.?	( ) Yes ( ) No	Could you give examples for "science diplomacy" approaches of not related to the EU or the U.S.?
According to your view, what would be necessary in order to develop coordinated transatlantic "science diplomacy" activities?	Please specify	

Annex 3: Organisations of Contributors to this Study

Type of organisations	Names of organisations
Universities and Research Organisations	• Austrian Institute of Technology, AT • George Mason University, U.S. • Georgetown University, U.S. • George Washington University/Elliot School, Business School, U.S. • Johns Hopkins University/School of Advanced International Studies, U.S. • Maryland University, U.S. • MIT Washington Office, U.S. • Salzburg University, AT
Think Tanks and Representative Organisations	• AAAS-American Association for the Advancement of Science (International, Security, Development, Policy), U.S. • Atlantic Council, U.S. • Brookings Institution, U.S. • Carnegy Endowment, U.S. • CSIS-Center for Strategic International Studies, U.S. • CRDF Global, U.S. • EARMA-European Association of Research Managers and Administrators • Fondation pour la recherche strategique, FR • Information Technology and Innovation Foundation, U.S. • ISC Intelligence, BE • National Academies of Sciences, U.S. • National Resources Defense Council, U.S. • OECD • Rand Corporation, U.S. • Transatlantic Academy, U.S. • U.S. Global Change Research Program, U.S. • World Bank • World Watch Institute, U.S.

(continued)

(continued)

Type of organisations	Names of organisations
Funding Organisations, Agencies	• Department of Commerce, U.S. • DoE-Department of Energy, U.S. • DoE-Department of Education, U.S. • DHS-Department of Homeland Security, U.S. • EPA-Environmental Protection Agency, U.S. • NIH-National Institutes of Health, U.S. • NSF-National Science Foundation, U.S. • National Nanotechnology Initiative, U.S. • USAID, U.S. • Austrian Basic Research Fund, AT • Bertelsmann Foundation, DE • Humboldt Foundation, DE
Individual states represented by science counsellors, ministry representatives, ministry contracted organisations	• Austria, Germany, India, Italy, Poland, Spain, Switzerland, Sweden
European Institutions	• European Parliament • European Commission, Directorate General Research
U.S. Government Institutions	• U.S. Congress, Science committee • U.S. Department of State • OSTP-Office of Science Technology Policy at the White House, U.S. • U.S. Embassy in Austria

Annex 4: List of Reports and References

AAAS Forum on Science and Technology Policy, S&T Challenges, Initiatives, and Budgets under tightening Fiscal Constraints, Presentation by John P. Holdren, Washington DC, 5 May 2011

AAAS Report XXXVI: Research & Development FY 2012

Addressing the Innovation Imperative, Seminar, Washington 4 Feb 210, Charles Wessner/National Academies of Sciences, Presentation

A Digital Agenda for Europe; European Commission COM (2010) 245 final, 2010

A more research-intensive and integrated European Research Area: Science, Technology and Competitiveness key figures report 2008/2009, European Commission, 2008

After America: Narratives for the next Global Age; Paul Starobin, Viking, 2009

A focus on competitiveness restructuring policymaking for results, John Podesta, Sarah Rosen Wartell, and Jitinder Kohli, Center for American Progress, December 2010

Alliance Reborn: An Atlantic compact for the twenty-first Century, Washington, DC: Washington NATO Project, 2009

America and the World: Conversations on the future of American Foreign Policy, Zbigniew Brzezinski, Brent Scowcroft, moderated by David Ignatius, Basic Books, 2008

America Competes Reauthorization Act, 2010

Americans are from Mars, Europeans are from Venus, R. Kagan, of paradise and power: America and Europe in the new world order, 2003

A strategy for American innovation: Driving towards sustainable growth and quality jobs, 2011

Atlantic century: Benchmarking EU & U.S. innovation and competitiveness, ITIF-Information Technology & Innovation Foundation, February 2009

Atlantic century II: Benchmarking EU & U.S. innovation and competitiveness, ITIF-information technology & innovation foundation, July 2011

Building the twenty-first Century: U.S. China Cooperation on Science, Technology, and Innovations; Charles W. Wessner, Rapporteur; Committee on Comparative National Innovation Policies: Best Practice for the twenty-first Century; National Research Council, 2011

Catalyzing American Ingenuity: The Role of Government in Energy Innovation, American Energy Innovation Council, 2011

China: Fragile Superpower; Susan L. Shirk, Oxford University Press, 2007, 2008

Coping with a Conflicted China; David Shambaugh, The Washington Quarterly, Winter 2011

Does America Need a Foreign Policy? Toward a Diplomacy for the 21st Century; Henry Kissinger, Simon & Schuster, 2001

Europe 2020: A strategy for smart, sustainable and inclusive growth; European Commission COM (2010) 2020, 2010

Energy 2020: A strategy for competitive, sustainable and secure energy; European Commission, COM (2010) 639 final, 2010

Europe 2020 Flagship Initiative Innovation Union; European Commission, COM (2010) 546 final, 2010

Europe 2020, Competitive or Complacent? Daniel S. Hamilton, 2011

European Defense Trends: Budgets, Regulatory Frameworks, and the Industrial Base, Report of the CSIS Defense-Industrial Initiatives Group, 2010

Europe Faces Outward, Responses to Globalization, U.S. policies for India and China, Woodrow Wilson center, European Alumni Association, conference 2008

EU Internal Security Strategy in Action: Five steps towards a more secure Europe, European Commission, COM (2010) 673 final, 2010

EU-U.S. Summit, Joint Statement, Lisbon 20 November 2010

Examination of the U.S. Air Force's Science, Technology, Engineering, and Mathematics (STEM) Workforce needs in the future and its strategy to meet those needs; National Research Council, 2010

Foreign Policy and Development: Structure, Process, Policy, and the Drip-by-Drip Erosion of USAID, Gerald F. Hyman, CSIS-Center for Strategic & International Studies, 2010

From the Lisbon Treaty to the Eurozone crisis: a new beginning or the unraveling of Europe? Summary of the 2010 cuse annual conference, Washington, D.C., June 2, 2010

GIS and Spatial Agent-Based model: Simulations for sustainable development, Claudio Cioffi-Revilla, J. Daniel Rogers, and Atesmachew Hailegiorgis, George Mason University, USA and Smithsonian National Museum of Natural History, USA, 2010

Global Forecast 2011, International Security in a Time of Global Uncertainty, CSIS-Center for Strategic and International Studies, 2011

Global Governance 2025: At a critical juncture, National Intelligence Council and the European Union's Institute for Security Studies, 2010

Hearing of John P. Holdren at house appropriations subcommittee on commerce, justice and science, 2011

Immigration Policy: Highly skilled workers and U.S. competitiveness and innovation; Darrell West, The Brookings Institution, 2011

Innovation Union Competitiveness Report 2011, European Commission

International Space Summit, Charles Bolden Statement, 17 Nov. 2010

Knowledge arbitrage, serendipity, and acquisition formality: their effects on sustainable entrepreneurial activity in regions, E. G. Carayannis, M. Provance, N. Givens, in IEEE Transactions on Engineering Management, 2011

Knowledge-driven creative destruction, or leveraging knowledge for competitive advantage, Strategic knowledge arbitrage and serendipity as real options drivers triggered by Co-opetition, co-evolution and co-specialization, E. G. Carayannis, 2008

Knowledge, Networks and Nations, Global Scientific Collaboration in the twenty-first Century, UK Royal Society 2011

Losing Control: The transatlantic partnership, the developing nations and the next phase of globalization, Joe Quinlan, Paper Series, Transatlantic Academy, March 2011

MASON RebeLand: An Agent-Based Model of Politics, Environment, and Insurgency, Claudio Cioffi-Revilla and Mark Rouleau; Center for Social Complexity, Krasnow Institute for Advanced Study, George Mason University, Fairfax, VA 22030, USA. Proceedings of the Human Behavior-Computational Modeling and Interoperability Conference 2009

Memorandum for the Heads of Executive Departments and Agencies, by the Executive Office of the President, Office of Management and Budget, 21 July 2010

National Preferences in publicly-supported R&D Programs, NEDO Report, G. R. Heaton, Jr., C. T. Hill, P. Windham, D.W. Cheney; May 2010

On China; Henry Kissinger, The Penguin Press, 2011

Persistent Forecasting of Disruptive Technologies Report 2, Committee on Forecasting Future Disruptive Technologies; National Research Council, 2010

President's council of advisors on science and technology (PCAST): Report to the President on ensuring American leadership in advanced manufacturing, June 2011

President's Council of Advisors on Science and Technology (PCAST): Report to the President on Accelerating the Pace of Change in Energy Technologies through an Integrated Federal Energy Policy, November 2010

President's Council of Advisors on Science and Technology (PCAST): Report to the President on Designing a Digital Future: Federally Funded Research and Development in Networking and Information Technology, December 2010

President Obama's Trip to India, Teresita C. Schaffer, CSIS-Center for Strategic & International Studies, 2010

Presidential Leadership to ensure Science and Technology in the Service of National Needs: A Report to the 2008 Candidates, Center for the Study of the Presidency Study Group on Presidential Science and Technology Personnel and Advisory Assets, 2008

Realising the New Renaissance, Policy proposals for developing a world-class research and innovation space in Europe 2030, Second Report of the European Research Area Board, 2010

Report on SFIC questionnaire: S&T cooperation with the United States of America, Dr. Ales Gnamus, knowledge for growth unit, IPTS, JRC, 2011

Rising above the Gathering Storm: Energizing and Employing America for a Brighter Economic Future, National Academies of Sciences, 2007

Rising Above the Gathering Storm, Revisited: Rapidly Approaching Category 5, by Members of the 2005 "Rising Above the Gathering Storm" Committee; Prepared for the Presidents of the National Academy of Sciences, National Academy of Engineering, and Institute of Medicine, 2010

Roadmap for Cybersecurity Research, Department of Homeland Security, Nov. 2009

Science and Engineering Indicators 2012, National Science Foundation/National Science Board (NSB 12–01), 2012

Science Diplomacy and Rhetoric-as-Epistemic: Finding Common Cause, M. Karen Walker, 2008

Shoulder to Shoulder, Forging a Strategic U.S.–EU Partnership, Daniel S. Hamilton (Editor), Washington, DC: Johns Hopkins University Center for Transatlantic Relations), 2010

Simulating European Union R&D policy—knowledge Dynamics in EU-funded Innovation Networks, Ramon Scholz, Terhi Nokkala, Andreas Pyka, Petra Ahrweiler and Nigel Gilbert, ESSA Conference 2009

STAR METRICS: Science and Technology for America's Reinvestment: Measuring the Effects of Research on Innovation, Competitiveness and Science: Stefano Bertuzzi, Office of Science Policy, Office of the Director, National Institutes of Health, Presentation at the First VIVO Annual Conference, August 13, 2010

State of the Union Address 2011 by U.S. President Obama, January 2011

State of the Union Address 2012 by U.S. President Obama, January 2012

S&T Strategies of 6 Countries: Implications for the United States Standing Committee on Technology Insight-Gauge, Evaluate & Review; National Research Council, 2010

Transatlantic Business Dialogue (TABD), several documents and reports

Transatlantic Economic Council (TEC), several documents and reports

The Agenda for the EU–U.S. strategic partnership, EU Institute for Security Studies, 2011

The Art and Politics of Science; Harold Varmus, W. W. Norton, 2009

The Atlantic Geopolitical Space: Common Opportunities and Challenges; Synthesis Report of a conference jointly organized by DG Research and Innovation and BEPA, European Commission, and held on 1 July 2011

The Business Plan: Our Commitment to Clean Energy, American Energy Innovation Council, 2010

The Dragon and the Elephant: Understanding the Development of Innovation Capacity in China and India: Summary of a Conference; Stephen Merrill, David Taylor, and Robert Poole, Rapporteurs; Committee on the Competitiveness and Workforce Needs of U.S. Industry; National Research Council, 2010

The Future Is Now: A Balanced Plan to Stabilize Public Debt and Promote Economic Growth; William A. Galston and Maya MacGuineas, Governance Studies at Brookings, 2010

The Future of Europe and Transatlantic Relations, Conference, Woodrow Wilson Center, 2009

The G20 Seoul Summit Leaders' Declaration, November 11–12, 2010

The Good, The Bad, and The Ugly (and The Self-Destructive) of Innovation Policy: A Policymaker's Guide to Crafting Effective Innovation Policy; Stephen J. Ezell and Robert D. Atkinson, ITIF-Information Technology and Innovation Foundation, 2010

The Innovator's Dilemma, When New Technologies Cause Great Firms to Fail, Clayton M. Christensen, Harvard Business School Press, 1997

The Lisbon Treaty in Focus: Germany, the EU, Transatlantic Relations, and Beyond, Frances G. Burwell, Ludger Kühnhardt, AICGS Policy Report, American Institute for Contemporary German Studies, Johns Hopkins University, 2010

The National Space Policy; The White House, 2010

The Origins of Political Order: From Prehuman Times to the French Revolution; Francis Fukuyama, Farrar, Straus and Giroux, 2011

The Power of Renewables: Opportunities and Challenges for China and the United States Committee on U.S.–China Cooperation on Electricity from Renewable Resources; National Research Council; Chinese Academy of Sciences; Chinese Academy of Engineering, 2010

The Science of Science Innovation Policy, Hearing before the Subcommittee on Research and Science Education Committee on science and Technology, House of Representatives, 23 September 2010

The Transatlantic Economy 2010, Hamilton, Daniel, and Quinlan, Joseph P., CTR-SAIS, Johns Hopkins University, 2010

The Transatlantic Economy 2011, Hamilton, Daniel, and Quinlan, Joseph P., CTR-SAIS, Johns Hopkins University, 2011

U.S. Government Engagement in International Cooperation in ST&I, John P. Holdren, Presentation at the Science Diplomats Club, Washington DC, December 2010

United States National Science Foundation: Investing in Americas Future, FY 2010, Agency Financial Report, 2010

Wider Europe: Unintended Strategic Consequences, CSIS-Center for Strategic & International Studies, 2010

Annex 5: List of Events

16 Nov 2010 New START treaty, Kurt Volker	Presentation, SAIS	http://www.csmonitor.com/Commentary/Opinion/2010/1116/Ratify-the-New-START-treaty-but-wait-until-January-to-do-it
18 Nov 2010 U.S. Foreign policy and USAID, USAID report by Gerald Hyman	Panel discussion, CSIS	http://csis.org/expert/gerald-hyman http://csis.org/event/foreign-policy-and-development-structure-process-policy
30 Nov 2010 Obamas Wars, Bob Woodward	Book presentation, discussion, George Washington University/Elliot School	http://media.elliott.gwu.edu/video/186
8 Dec 2010 Transatlantic Coop. in Homeland Security, EC Commissioner Cecilia Malström	Presentation SAIS	http://transatlantic.sais-jhu.edu/bin/g/s/12.8.10malmstrom.pdf
8 Dec 2010 New conservatism, justin Vaisse	Discussion, committee for economic development	http://www.ced.org/news-events/general/573-policy-luncheon-featuring-justin-vaisse
9 Dec 2010 Central Europe–time for leadership	Conference, SAIS	http://transatlantic.sais-jhu.edu/bin/s/y/12.9.10v4.pdf
9 Dec 2010 Foreign policy and development: Ruth Levine, Deputy assistant administrator Bureau for policy, planning and learning, USAID	Presentation, brookings	http://www.usaid.gov/press/speeches/2010/sp101209.html
9 Dec 2010 Business and social innovation,	Presentation, SAIS	http://www.sais-jhu.edu/calendar/index.htm svanlare@jhu.edu
10 Dec 2010 Transatlantic and Russia	Conference, SAIS	http://transatlantic.sais-jhu.edu/bin/u/h/12.10.10russia.pdf
15 Dec 2010 Energy innovation 2020	Conference, National Press Club	http://thebreakthrough.org/blog/2010/12/energy_innovation_2010_rethink.shtml
16 Dec 2010 EU security	Conference, U.S. Institute of Peace,	http://www.usip.org/events/the-eus-role-in-security-sector-reform
14 Dec 2010 Climate change discussion based on book by Roger Pielke	Panel discussion, Austrian embassy	http://www.ostina.org/content/view/5386/895/
12 Jan 2011 2011, Europe and the U.S., Werner Hoyer, DE Staatsminister im Auswärtigen Amt	Presentation, Friedrich Naumann Stiftung	http://aussen-sicherheitspolitik.de/?p=4491#more-4491
11 Jan 2011 Military—industry complex: Eisenhower farewell address	Panel discussion, New America foundation	http://www.newamerica.net/events/2011/military_industrial_complex
18 Jan 2011 Eisenhower farewell: about the role of scientists	Panel discussion, AAAS	http://www.cspo.org/documents/enews_10-dec.pdf
19 Jan 2011 Science diplomacy and nuclear	Conference, National Academies together with Institute of Peace	http://sites.nationalacademies.org/PGA/cisac/PGA_060004

(continued)

(continued)

31 Jan 2011 Joint committee meeting on atlantis program and fulbright	Half day meeting involving EC and DoS, DoEdu participants	
27 Jan 2011 DoS Diplomacy, Anne-Marie Slaughter	Presentation, SAIS	http://www.state.gov/s/p/rem/155622.htm
3 Feb 2011 Report on Cancun results, Jonathan pershing, deputy special envoy for climate change, DoS	Conference, SAIS	
4 Feb 2011 Wilson center, Foreign policy in 112th Congress	Panel discussion	http://www.wilsoncenter.org/index.cfm?fuseaction=events.event_summary&event_id=650348
7 Feb 2011 Cut or invest	Panel Discussion, ITIF	http://www.itif.org/events/cut-or-invest-what%E2%80%99s-best-way-grow-our-economy
7 Feb 2011 The future of power	Book presentation and panel discussion, CSIS	http://csis.org/event/book-launch-future-power
9 Feb 2011 EU–U.S. Energy cooperation and energy council	Panel discussion, Woodrow Wilson Center	http://www.wilsoncenter.org/index.cfm?fuseaction=events.welcome
15–16 Feb 2011 National science board meeting, NSF	Opportunity to listen to open parts, Universities report, etc.	http://www.nsf.gov/nsb/meetings/
17 Feb 2011 Janet Napolitano	Hearing at Senate committee on Homeland security and Governmental affairs	http://hsgac.senate.gov/public/index.cfm?FuseAction=Hearings.Hearing&Hearing_ID=17c95c54-71d4-4f58-b3bd-88d011e326b6
18–21 Feb 2011 AAAS annual meeting	Conference	
23 Feb 2011 Transatlantic security policy and discussion	Conference, CSIS	http://csis.org/event/enhancing-euro-atlantic-security-amid-uncertain-times-eu-us-security-strategies-and-recommenda
3 Mar 2011 Future of U.S. research Universities, Provost Johns Hopkins University	Presentation, SAIS	http://www.sais-jhu.edu/bin/m/x/provost.pdf http://www.sais-jhu.edu/events/spring2011/minor.htm
9 Mar 2011 The Obama administration's Innovation policy	Conference, ITIF	http://www.itif.org/events/obama-administrations-innovation-policy
10 Mar 2011 U.S. Competitiveness: A new conversation with new opportunities	Conference, ITIF	http://www.itif.org/events/us-competitiveness-new-conversation-new-opportunities
10–11 Mar 2011 U.S.–EU bridging Nano EHS research efforts	Conference, Nano initiative	Nano Strategy 2011: http://www.nano.gov/nnistrategicplan211.pdf http://www.nano.gov/html/meetings/us-eu/index.html
29 Mar 2011 New University of the twenty-first Century, Western Governors University	Presentation, center for Economic development	http://www.ced.org/news-events/postsecondary-education/629-ced-policy-luncheon-on-post-secondary-education-with-dr-robert-mendenhall-president-and-ceo-western-governors-university
30–31 Mar 2011, Catastrophes and complex systems: transportation, Department of Homeland security science conference	Conference, Department of Homeland security	http://www.orau.gov/dhssummit/agenda.htm

(continued)

(continued)

8 April 2011 Transatlantic climate security	Conference, Carnegie Endowment	http://www.carnegieendowment.org/events/?fa=eventDetail&id=3204
12 April 2011 U.S. budget with Chis van Hollen, center for American programm	Presentation, center for American progress	http://www.americanprogressaction.org/events/2011/04/fedbudget.html
20 April 2011 Towards a new Architecture for politico-military security in Europe	Conference, SAIS	http://www.gwu.edu/~ieresgwu/assets/docs/4.18.11.pdf http://transatlantic.sais-jhu.edu/bin/a/b/4.21.11osce.pdf
20 April 2011 commercial imperialism of the U.S. during the cold War	Conference, SAIS	http://aidwatchers.com/2010/05/commercial-imperialism-political-influence-and-trade-during-the-cold-war/
2 May 2011 Barney Frank	Presentation, Center for American Progress	http://www.defensenews.com/story.php?i=6393461&c=AME&s=BUD http://www.americanprogressaction.org/pressroom/2011/04/sustainable_defense_posture.html
4 May 2011 John P. Holdren	Hearing, house Appropriations commerce, justice, science subcommittee	http://appropriations.house.gov/_files/AppropriationsCommitteeWeeklyHearingSchedule52to56.pdf http://www.spacenews.com/policy/100225-house-appropriators-grill-obama-science-adviser.html http://www.msnbc.msn.com/id/42934529/ns/technology_and_science-space/t/obama-sees-china-partner-mars-mission/
4 May 2011 John P. Holdren	Lecture, George Washington University/Elliot School	http://www.gwu.edu/~cistp/ http://media.elliott.gwu.edu/video/229
5–6 May 2011 AAAS science policy forum 2011	Conference	https://www.signup4.net/Public/ap.aspx?EID=STPO10E&OID=50 John P. Holdrens Position: http://www.aaas.org/news/releases/2011/0506stpf_holdren.shtml

Annex 6: Abbreviations

Abbreviation	Explanation
BRIC	Brazil, Russia, India, China
DG	Directorate General
EC	European Commission
EU	European Union
FDI	Foreign Direct Investment
ITAR	International Traffic in Arms Regulations
NGO	Non Governmental Organization
R&D	Research and Development
S&T	Science and Technology
TABD	Transatlantic Economic Business Dialogue
TEC	Transatlantic Economic Council
U.S.	United States of America

Index

S. E. Herlitschka, *Transatlantic Science and Technology*, SpringerBriefs in Business,
DOI: 10.1007/978-1-4614-4385-8, © Sabine E. Herlitschka 2013

Printed by Publishers' Graphics LLC
BT20121017.19.19.42